GWM-2005—A Groundwater-Management Process for MODFLOW-2005 with Local Grid Refinement (LGR) Capability

By David P. Ahlfeld[1], Kristine M. Baker[1], and Paul M. Barlow[2]

[1] University of Massachusetts.

[2] U.S. Geological Survey.

Groundwater Resources Program

Techniques and Methods 6–A33

U.S. Department of the Interior
U.S. Geological Survey

U.S. Department of the Interior
KEN SALAZAR, Secretary

U.S. Geological Survey
Marcia K. McNutt, Director

U.S. Geological Survey, Reston, Virginia: 2009

For more information on the USGS—the Federal source for science about the Earth, its natural and living resources, natural hazards, and the environment, visit http://www.usgs.gov or call 1-888-ASK-USGS

For an overview of USGS information products, including maps, imagery, and publications, visit http://www.usgs.gov/pubprod

To order this and other USGS information products, visit http://store.usgs.gov

Suggested citation:
Ahlfeld, D.P., Baker, K.M., and Barlow, P.M., 2009, GWM-2005—A Groundwater-Management Process for MODFLOW-2005 with Local Grid Refinement (LGR) capability: U.S. Geological Survey Techniques and Methods 6–A33, 65 p.

Preface

This report describes an updated version of the Groundwater-Management (GWM) Process for the 2005 version of the U.S. Geological Survey modular three-dimensional groundwater model, MODFLOW-2005. The performance of the program has been tested in a variety of applications, some of which are documented in this report. Future applications, however, might reveal errors that were not detected in the test simulations. Users are requested to notify the U.S. Geological Survey of any errors found in this report or the computer program by using the address on the inside of the back cover of the report. Updates might occasionally be made to both the report and to the computer program. Users can check for updates on the Internet at URL http://water.usgs.gov/software/lists/groundwater/.

Contents

Figures

Tables

Conversion Factors

Multiply	By	To obtain
cubic foot (ft^3)	0.02832	cubic meter (m^3)
cubic foot per day (ft^3/d)	0.02832	cubic meter per day (m^3/d)
foot (ft)	0.3048	meter (m)
foot per day (ft/d)	0.3048	meter per day (m/d)
square foot per day (ft^2/d)	0.09290	square meter per day (m^2/d)

Abbreviations

Symbol	Explanation
GWF	Groundwater Flow (Process)
GWM	Groundwater Management
LGR	Local Grid Refinement
LP	linear programming
MPS	Mathematical Programming System
OBS	Observation (Process)
SEN	Sensitivity (Process)
SLP	sequential linear programming

GWM-2005—A Groundwater-Management Process for MODFLOW-2005 with Local Grid Refinement (LGR) Capability

By David P. Ahlfeld, Kristine M. Baker, and Paul M. Barlow

Abstract

This report describes the Groundwater-Management (GWM) Process for MODFLOW-2005, the 2005 version of the U.S. Geological Survey modular three-dimensional groundwater model. GWM can solve a broad range of groundwater-management problems by combined use of simulation- and optimization-modeling techniques. These problems include limiting groundwater-level declines or streamflow depletions, managing groundwater withdrawals, and conjunctively using groundwater and surface-water resources. GWM was initially released for the 2000 version of MODFLOW.

Several modifications and enhancements have been made to GWM since its initial release to increase the scope of the program's capabilities and to improve its operation and reporting of results. The new code, which is called GWM-2005, also was designed to support the local grid refinement capability of MODFLOW-2005. Local grid refinement allows for the simulation of one or more higher resolution local grids (referred to as child models) within a coarser grid parent model. Local grid refinement is often needed to improve simulation accuracy in regions where hydraulic gradients change substantially over short distances or in areas requiring detailed representation of aquifer heterogeneity. GWM-2005 can be used to formulate and solve groundwater-management problems that include components in both parent and child models. Although local grid refinement increases simulation accuracy, it can also substantially increase simulation run times.

Introduction

The Groundwater-Management (GWM) Process was developed for the U.S. Geological Survey modular three-dimensional groundwater model (MODFLOW) to solve a broad range of steady-state or transient groundwater-management problems by combined use of simulation- and optimization-modeling techniques (Ahlfeld and others, 2005). Such problems include limiting groundwater-level declines or streamflow depletions, managing groundwater withdrawals, and conjunctively using groundwater and surface-water resources. The simulation-optimization modeling approach combines a simulation model of a groundwater system with an optimization (or management) model of the groundwater-management problem to be solved. Most often, numerical models are used to simulate the groundwater system; in this work, the numerical code that is used is the Groundwater Flow (GWF) Process of MODFLOW (Harbaugh and others, 2000; Harbaugh, 2005).

Formulation of an optimization model consists of three components—a set of decision variables, a management objective, and a set of management constraints. GWM was designed to support several types of decision variables and constraints that are commonly used in groundwater problems. Examples of decision variables supported by GWM include withdrawal or injection rates at managed wells, application rates at artificial-recharge basins, and interbasin transfers of water into or out of a groundwater basin. Management constraints supported by GWM include upper and lower bounds on pumping and injection rates, water-supply demands, hydraulic-head constraints such as drawdowns and hydraulic gradients, and streamflow and streamflow-depletion constraints. A single type of objective function currently (2009) can be specified in GWM, which is to maximize or minimize the sum of decision variables specified in the management formulation. The resulting management model is solved by GWM by use of a widely applied simulation-optimization modeling technique called the response-matrix approach, which is described in detail in Ahlfeld and others (2005).

The original GWM Process was developed for the 2000 version of MODFLOW (called MODFLOW-2000; Harbaugh and others, 2000). Since publication of GWM, a new version of MODFLOW has been released (MODFLOW-2005; Harbaugh, 2005). This new version of MODFLOW is similar in overall design to MODFLOW-2000, but incorporates a different approach for managing internal data (Harbaugh, 2005). One of the reasons for developing MODFLOW-2005 was to provide a programming structure to facilitate simulation of multiple model grids within a single model domain. Simulations of groundwater flow and transport often need highly refined grids to improve simulation accuracy in local areas of interest within a larger, regional-scale flow system. For example, locally refined grids may be needed in regions where hydraulic gradients change substantially over short distances, in areas requiring detailed representation of aquifer heterogeneity, or in regions of site-scale contamination where simulations of plume movement are of interest (Mehl and Hill, 2005). Mehl and Hill (2005 and 2007) developed a local grid refinement (LGR) capability for MODFLOW-2005 to provide a means to simulate higher resolution local grids (referred to as child models) within a coarser grid parent model. LGR accomplishes this by use of a shared-node, iterative-coupling approach, in which heads and fluxes are balanced across the shared interfacing boundaries of the parent and child models. The function of the child model is to simulate phenomena that need a finer grid than is provided by the parent model; the role of the parent model is to provide boundary conditions that are consistent with the larger scale flow system for the child model (Mehl and Hill, 2005). An example of a coarse-grid parent model with two areas of local grid refinement (that is, two child models) is illustrated in figure 1.

It is anticipated that MODFLOW-2005 will become the standard MODFLOW version in use; because there are situations in which the management of groundwater resources is of interest at both local and regional scales, it is desirable to integrate the local grid refinement capability of MODFLOW into the GWM Process. Consequently, the GWM Process has been updated to be compatible with MODFLOW-2005 and its local grid refinement capability.

EXPLANATION

- Node of the parent model only
- Shared node used by both the parent model and the child model
- Node of the child model only. The parent model is inactivated here after the initial parent simulation, so the parent model has a hole in it.
- Specified-head boundary node of the child model determined by interpolation from the parent solution at the shared nodes

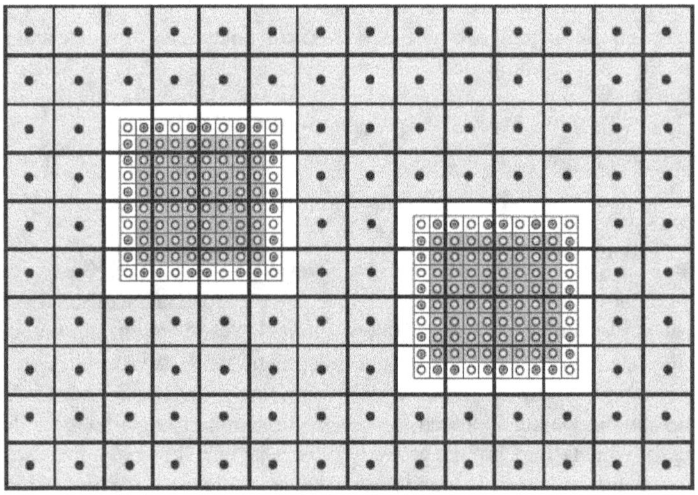

Figure 1.   Schematic diagram of a coarse grid with two areas of local grid refinement separated by two parent grid cells. Figure modified from Mehl and Hill (2007).

Purpose, Scope, and Terminology

The purpose of this report is to describe the new Groundwater-Management Process for MODFLOW-2005, which includes the capability to simulate multiple model grids by use of MODFLOW's LGR process. The report also describes several modifications and enhancements that have been made since the initial release of the GWM Process. The report includes revised input instructions for the GWM Process and three sample problems that demonstrate the combined use of GWM and LGR capabilities.

Throughout this report, the original GWM Process is referred to as GWM-2000, because it is based on the 2000 version of MODFLOW; in contrast, the updated GWM Process is referred to as GWM-2005, because it is based on MODFLOW-2005. When discussion in the text is applicable to either the GWM-2000 or GWM-2005 versions of the code, the general term GWM is used. Likewise, when discussion in the text is applicable to either the 2000 or 2005 versions of MODFLOW, the general term MODFLOW is used.

GWM-2005 can be used in one of four modes:

- MODFLOW-2005 GWF Process without GWM or LGR capabilities,

- MODFLOW-2005 GWF Process with LGR capability but without GWM,

- GWM with the MODFLOW-2005 GWF Process but without LGR, and

- GWM with the MODFLOW-2005 GWF Process with the LGR capability.

For users who will not need the LGR capability, the revised GWM Process works much as the original, and the user can skip those sections of the report that are concerned with LGR. The MODFLOW-2005 manual (Harbaugh, 2005), the original GWM documentation report (Ahlfeld and others, 2005), and the reports describing the LGR capability by Mehl and Hill (2005 and 2007) should be used as companions to this report.

Throughout the report, all file types are shown in bold uppercase text, such as MODFLOW's **NAME** file. Keywords used in MODFLOW, LGR, and GWM input files are shown in bold uppercase text in italics, such as the keyword ***DATA***, which is used to identify a particular file type.

Groundwater-Management (GWM) Process for MODFLOW

GWM links MODFLOW with a set of optimization-modeling techniques to solve several types of linear, nonlinear, and mixed-binary linear groundwater-management problems. It is presumed that most readers of this report are familiar with the development, calibration, and use of groundwater models based on the MODFLOW simulation code, but probably less familiar with the development and application of an optimization (or management) model. Therefore, the emphasis in this discussion is on the several components of an optimization model and how they are implemented in GWM. Readers are referred to Ahlfeld and others (2005) for a complete description of the theory, implementation, and use of the GWM Process.

An optimization model consists of two primary phases: a formulation phase and a solution phase. The formulation phase consists of defining a set of decision variables, an objective function, and a set of constraints that compose the management problem. The solution phase uses one of several solution techniques to solve the management problem. The discussion that follows describes the components of the formulation and solution phases that have been implemented in GWM, including a description of the files that are used in a GWM simulation.

Several required and optional input files are used to define a management formulation in GWM. The first of these is the **DECVAR** file, which is a required file that is used to define the decision variables of the management problem. The decision variables of a management problem are the quantifiable controls (or decisions) that are to be determined by the optimization model, such as the withdrawal rates from a set of managed wells. Currently, GWM supports three types of decision variables: flow-rate decision variables, external decision variables, and binary variables. Of these three, the most widely used are flow-rate decision variables, which consist of a withdrawal (discharge) or injection (recharge) flow rate at model cells. Each flow-rate decision variable can extend over one or more model cells and can be active during one or more stress periods. For example, a flow-rate decision variable might represent a steady-state application rate to an artificial-recharge basin that extends over multiple model cells, or a single withdrawal rate that represents total withdrawal from a wellfield composed of five wells distributed over a multicell area. The model cells defined for a flow-rate decision variable need not be adjacent to one another or even in the same model layer. Flow rates defined as decision variables are referred to as managed flows, because they are part of the management problem. The user can also specify unmanaged flows in a GWM simulation by use of the Well or Multi-Node Well (Halford and Hanson, 2002) Packages of MODFLOW. Unmanaged flows are not part of the management formulation and therefore are not changed during the GWM solution process.

External decision variables are general-purpose variables that do not have a direct effect on the state variables of the flow system (that is, heads, groundwater-flow rates, streamflow rates, and so forth). A typical use of an external variable is to represent water that is imported to a groundwater basin from a distant surface-water reservoir. This water might be used, for example, to supplement the water-supply demand of a basin that cannot be met by withdrawals from aquifers within the basin. An example of the use of external variables to represent surface-water reservoir storage and river flows for a conjunctive-use water-management problem is provided by Pulido-Velazquez and others (2008). Because external variables are commonly used to represent flows of water, external variables are referred to in GWM as either imports or exports.

Binary variables are a special type of decision variable that can have only a value of zero or one. Binary variables are used to define the status of the flow-rate and (or) external variables that have been associated with each binary variable. For example, if a binary variable has been associated with a particular managed well, the binary variable takes a value of one if the managed well is active in the management solution and a value of zero if the managed well is inactive in the solution. Any combination of flow-rate or external variables can be assigned to a single binary variable. For example, a binary variable might be associated with 12 flow-rate decision variables that represent the 12 monthly withdrawal rates from a single well. If the well is pumped in any of the 12 months, then the binary variable takes a value of one. Binary variables are often used to model construction costs of a water-supply facility. With reference to the previous example, if pumping from the well occurs in any of the 12 months, then the well must be constructed, which would incur a cost to the operation of the water-supply system.

The objective function of a management problem is used to identify the best solution among many possible solutions to a particular management problem. The objective function is stated in terms of one or more of the decision variables defined in a management problem. Currently, GWM supports a single type of objective function, which is a weighted linear summation of the three types of decision variables. The weights consist of cost and benefit coefficients and, for the flow-rate and external variables, the total duration of time that each decision variable is active. The weighted linear summation can be either maximized or minimized. The objective function of a GWM problem is defined in the **OBJFNC** file, which is also a required input file.

The constraints of a management problem impose restrictions on the values that can be taken by the decision variables. Currently, GWM supports four types of constraints. The first are called decision-variable constraints, and are the upper and lower bounds on the values that each of the flow-rate and external variables can take in the solution. For example, these constraints can be used to define the maximum and minimum withdrawal rates from a particular managed well. Decision-variable constraints are specified in the **VARCON** file, which is a required input file.

The second constraint type is called a linear-summation constraint. This type of constraint makes it possible to define weighted linear summations of any of the three types of decision variables, where the weights are scalar coefficients. The sum of the decision variables can be greater than, less than, or equal to a specified value. Examples of the use of linear-summation constraints are (1) that the sum of withdrawals from a set of wells must be greater than or equal to a water-supply demand, (2) that the amount of water imported from an external surface-water source must be less than a specified amount, or (3) that the total number of wells in a water-supply network must be less than a specified maximum value. Linear-summation constraints are specified in the **SUMCON** file, which is an optional input file.

The third and fourth types of constraints are based on model-calculated state variables of the flow system, specifically, hydraulic heads and streamflows. Hydraulic-head constraints are specified in the **HEDCON** file, and streamflow constraints are specified in the **STRMCON** file, both of which are optional files. Four types of hydraulic-head constraints can be specified. The first are lower and (or) upper bounds on heads defined for specific model cells and stress periods. The second are drawdowns defined for specific model cells and stress periods. The third and fourth types are based on the head difference between two model cells for a specific stress period. The first of these is a lower bound placed on the difference in head between the two cells. Similarly, a constraint can be placed on the lower bound of the gradient in head between two cells, which is simply the difference in head between the two cells divided by the distance between the cells.

Two types of streamflow constraints can be specified in the **STRMCON** file: upper and (or) lower bounds on streamflow at a specific stream location and stress period and upper and (or) lower bounds on streamflow depletion at a specific stream location and stress period. Streamflow depletion is defined as the difference in model-calculated streamflow between an initial streamflow value and that calculated after implementation of the optimal management strategy. Streamflow can be simulated with either the STR (Prudic, 1989) or SFR (Prudic and others, 2004; Niswonger and Prudic, 2005) Streamflow-Routing Packages of MODFLOW (described in greater detail below).

GWM can be used to solve management problems that are formulated as linear programs, nonlinear programs, or mixed binary programs. A linear program is an optimization formulation in which the objective function and all of the constraints are linear functions of the decision variables. A nonlinear program is one in which the objective function and (or) one or more of the constraints are nonlinear functions of the decision variables. Nonlinear programs are necessary to solve groundwater-management problems for conditions that include water-table (unconfined) aquifers or head-dependent boundary conditions such as streams, drains, or evapotranspiration from the water table. A mixed binary program is a formulation that includes binary decision variables.

The basis of the solution techniques that are used by GWM to solve each of the three types of formulations is called the response-matrix approach, which has been widely applied in groundwater simulation-optimization modeling. The response-matrix approach is based on the determination of response coefficients, which are the ratio of the change in the value of head or streamflow at each constraint location to a perturbation (change) in the value of a flow-rate decision variable, such as the withdrawal rate at a well. GWM uses the GWF Process of MODFLOW to calculate response coefficients. For linear formulations, the resulting matrix of response coefficients is then combined with other components of the linear-management formulation to form a complete linear formulation; the formulation is then solved by GWM by use of an optimization technique called the simplex algorithm (Dantzig, 1963; Gass, 1985). Nonlinear formulations are solved by sequential linear programming (SLP); that is, repeated linearization of the nonlinear features of the management problem. In this approach, response coefficients are recalculated in each iteration of the solution process. An example application of the SLP algorithm to a field-scale, unconfined-aquifer management problem is provided by Ahlfeld and Baro-Montes (2008). Mixed-binary formulations are solved by use of the branch and bound algorithm (Nemhauser and Wolsey, 1988) in conjunction with the response-matrix approach. It should be noted that nonlinear and mixed-binary formulations can add significant computational burden to the GWM solution process. In addition, because the branch and bound algorithm is designed to operate with a linear program, the use of GWM to solve problems that include both nonlinear elements and binary variables may not yield the global optimal solution (see Ahlfeld and Mulligan, 2000, p. 147–149, for a discussion of global solutions for nonlinear problems). Information about the particular type of solution process to be used in a GWM run is specified in the **SOLN** file, which is a required input file.

Local Grid Refinement Capability of MODFLOW-2005

The local grid refinement option of MODFLOW-2005 provides the capability to simulate one to nine block-shaped, higher resolution local grids (child models) within a coarser grid (parent model). The refined areas cannot overlap, and at least one parent cell must separate refined areas (Mehl and Hill, 2007). The LGR capability can be used in two- or three-dimensional steady-state or transient simulations and in simulations of confined and (or) unconfined groundwater systems. Mehl and Hill (2005 and 2007) provide detailed explanations of the theory and development of the LGR capability, as well as discussions of the accuracy and execution times of LGR simulations. The discussion below briefly describes the shared-node LGR algorithm and highlights some of the important issues that need to be considered for implementation of the LGR capability.

Figure 1 illustrates a coarse-grid parent model with two areas of local grid refinement (that is, two child models). Cells fully within the child grids are darkly shaded in the figure, whereas cells fully within the parent grid are lightly shaded and cells at the interface are white. Each parent and child model joins along what are called interfaces. The interior cells of the parent model that are covered by the child models are made inactive by LGR by setting variable IBOUND in the MODFLOW Basic Package to zero. The open circles along the interfaces between the child and parent models in figure 1 are the shared nodes used by both the parent and child models.

The following list summarizes some of the issues that must be considered in the grid and time-step discretization of the parent and child models:

- The lateral boundaries of the child models are always interfaces with active cells of the parent model. That is, there must be at least one row or column of active cells of the parent model between each child model and the lateral extent of the active cells of the parent model.

- The shared-node coupling used by LGR requires child-grid spacing along rows and columns that is an odd-integer multiple of the parent grid (3:1, 5:1, and so forth), and the refinement ratio along rows and columns needs to be the same in the two directions for all rows and columns.

- For vertical refinement, the bottom of the child-model grid coincides with the finite-difference nodes in any layer of the parent-model grid except in the top and bottom layers of the parent model (figs. 2A-C). That is, except for the top and bottom layers of the parent model, the bottom layer of the child model replaces one half of the thickness of the parent cell (for example, fig. 2B). For vertical refinement within the bottom parent layer, the extent of the child grid needs to coincide with the bottom of the parent grid (fig. 2C). For vertical refinement of parent models with a single model layer, the child model extends from the top to the bottom of the layer (fig. 2A).

- Vertical refinement can differ layer by layer (see, for example, figures 2B and 2C), but child-grid spacing in the vertical dimension also must be an odd-integer multiple of the parent grid. Moreover, for vertical refinement, the top of the child model needs to coincide with the top of the parent model. In other words, the top boundary of the child model is always the top of the simulated saturated groundwater system of the parent model; however, vertical grid refinement does not need to start at the top because a vertical refinement ratio of 1:1 can be used in the top layer (fig. 2C). This 1:1 ratio can be useful when thick upper layers are desired for simulating water-table conditions.

- The simulation of unconfined aquifers by the LGR algorithm can cause problems when the rewetting capability of MODFLOW is used (Mehl and Hill, 2005, p. 19). The child grids need to be constructed so that they will not be adjacent to areas of the parent grid that are likely to go dry. Also, the use of thicker upper layers in the child model can be advantageous in alleviating some of the drying and rewetting problems associated with thin layers at the top of a model.

- For transient simulations, LGR requires that the parent and child models use equivalent time discretization. This is most easily accomplished by defining identical stress-period lengths and time-step variables (input variables PERLEN, NSTP, and TSMULT) in the MODFLOW Discretization input file.

LGR uses an algorithm based on a two-way iterative coupling of the parent and child models. The approach iteratively updates the head and flux boundary conditions along the interface for each model. The algorithm begins with a solution of the parent model that encompasses the entire simulation domain and ignores each of the child models. For subsequent iterations, the parent-model cells that are completely covered by the refined areas are inactivated; thus, after the initial parent-model solution, the parent model has one or more holes that are filled by each child model. The child models are then simulated by setting specified-head boundary conditions at the shared nodes equal to the heads calculated at these nodes from the parent-model simulation; for child-grid nodes along the boundary that do not coincide with a parent node (the nodes represented by double circles in figure 1), the specified head is interpolated by methods described in Mehl and Hill (2005, p. 15–19). Each child model is simulated before resimulating the parent model (Mehl and Hill, 2007). That is, the effects of all child models are accumulated and applied to the parent model in a single parent simulation. In this approach, the ordering of the child-model simulations does not affect the convergence or the results.

The initial sequence of simulation of the parent model followed by simulation of each of the child models completes the first iteration of the LGR coupled-model algorithm. Subsequent iterations of parent-child models use specified-flux boundary conditions along the parent-grid interface that are calculated by use of a mass balance on the child cells that border each parent cell. Relaxing (averaging) with the head and flux values from the previous iteration is needed for convergence and the stability of subsequent iterations (Mehl and Hill, 2005, p. 13–15). Two relaxation variables must be specified by the user in the LGR **CONTROL** file for each parent-child model pair: variable RELAXH, the relaxation factor for heads, and variable RELAXF, the relaxation factor for fluxes. The value of the relaxation parameters is problem dependent and may need to be adjusted to achieve convergence.

Iteration between the parent and child models continues until two simulation closure criteria have been met—one for the parent model and one for the child model. For convergence of the parent model, the maximum relative change of the coupling specified-flux boundary condition between successive iterations needs to be less than a user-defined amount; for convergence of the child model, the maximum change of the coupling specified-head boundary condition between successive iterations needs to be less than a user-defined amount. The closure criteria are specified by use of variables FCLOSELGR and HCLOSELGR in the LGR CONTROL file (described in the "Input Instructions and Output Files" section of this report); separate sets of closure criteria are specified for each parent-child model pair. The user also specifies the maximum number of LGR iterations that are allowed before the iterative process is stopped (variable MXLGRITER in the LGR CONTROL file).

A second level of convergence also needs to be considered for each LGR simulation, namely, convergence of each of the individual parent and child flow models during each LGR iteration. The accuracy of each of the parent and child models is controlled by convergence closure criteria that are set in the MODFLOW solver packages (SIP, PCG2, DE4, and GMG) for heads and (or) residual flows. The parent and each child model can use different solvers. Generally, the closure criteria used for the solvers should be less than or equal to what is used for the LGR closure (Mehl and Hill, 2005, p. 20). As noted by Mehl and Hill, it does not make sense to try to solve the LGR coupling boundary conditions to a greater precision than the closure criteria specified for each of the individual models. Most of the solver packages will iterate until one or both of the convergence criteria are met or until a maximum number of iterations is reached. Successful LGR convergence does not necessarily imply that the parent and child models have also successfully converged. It is possible, for example, that a child model does not converge but instead iterates until the maximum number of iterations allowed is reached. As long as the fluxes and heads at the perimeter of the child grid are unchanged from a prior iteration, LGR will consider convergence to have been achieved, but a statement that the flow process did not converge will be written to the output file. To avoid possible problems that this unlikely scenario might cause, a set of tests for achievement of convergence at both the individual-model and LGR solution levels is implemented in GWM-2005.

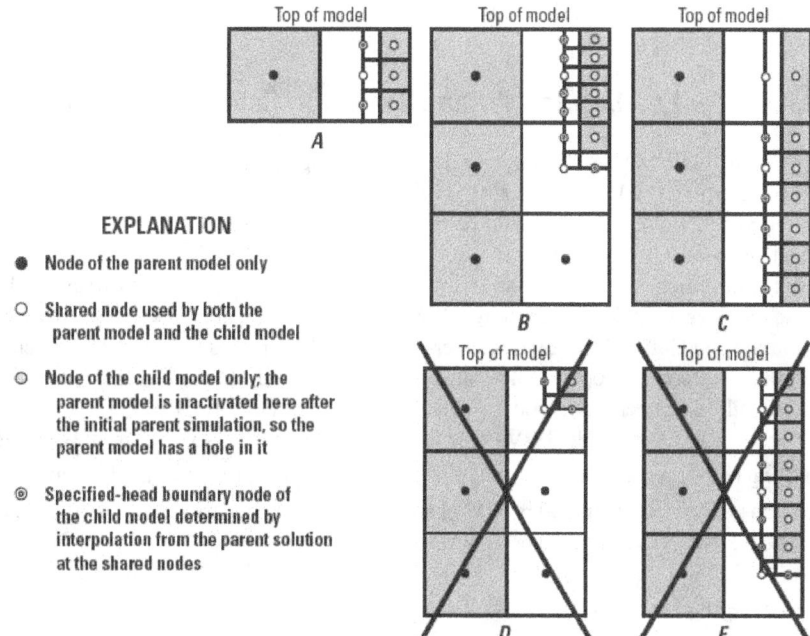

EXPLANATION

● Node of the parent model only

○ Shared node used by both the parent model and the child model

○ Node of the child model only; the parent model is inactivated here after the initial parent simulation, so the parent model has a hole in it

⊚ Specified-head boundary node of the child model determined by interpolation from the parent solution at the shared nodes

Figure 2. Cross-sectional schematic diagram of a of vertical-refinement interface of *(A)* a one-layer parent model refined to a three-layer child model, *(B)* a multilayer parent model in which the child refinement varies vertically and terminates at the shared node of the second parent layer, *(C)* a multilayer parent model in which the child refinement varies vertically and extends to the bottom of the parent model, *(D)* a multilayer parent model in which the child refinement terminates at the first shared node of the parent, which is not allowed, and *(E)* a multilayer parent model in which the child refinement terminates at the bottommost shared node, which is not allowed. Figure from Mehl and Hill (2005).

New Features in GWM

Several modifications and enhancements have been made to GWM since its initial release as GWM-2000. The major changes are summarized here.

These features were added to GWM-2000 and included in GWM-2005:

- Writing the response matrix as a formatted output file,

- Turning off the automated reexecution of the GWF Process,

- Writing optimal flow rates in the **WEL**-file format,

- Providing additional output at each SLP iteration, and

- Using the GMG solver (Wilson and Naff, 2004) for the GWF Process.

These features are available only in GWM-2005:

- Directing output from GWM to a named GWM **OUT** file instead of the **GLOBAL** file,

- Writing the response matrix to a formatted file for the NS (no solution) option,

- Specifying streamflow constraints with the SFR Package,

- Providing an option to adjust the head closure criteria for GWF Process runs, and

- Formulating optimization problems on multiple models with LGR.

With the exception of the LGR capabilities, which are described in the next section of this report, these new features are discussed in this section or in the relevant input-file descriptions. Major modifications to the GWM-2000 computer code that were implemented to conform to the structure of MODFLOW-2005 and to accommodate these new features are described in the Appendix ("Program Modifications to Convert GWM-2000 to GWM-2005").

Enhanced Options for Reading and Writing Response-Coefficient Matrixes

The initial version of GWM provided three options for reading and writing a response-coefficient matrix when the linear programming (LP) solution option is used. These three options did not allow the response matrix to be written to a formatted file that could be easily viewed by a model user. However, because of the importance of the response coefficients to the solution of an optimization model as well as to the analysis of optimization results, the options for reading and writing response matrixes have been expanded. The method by which response matrixes are read or written by GWM depends on the value of variable IRM specified in item 4a of the Solution and Output-Control Parameters (**SOLN**) file. Options 0, 1, and 2 are unchanged from the original GWM release, and new options 3, 4, and 5 have been added. (In the discussion that follows, the term "nonformatted" file is a general description. The specific type of nonformatted file is designated in the *openspec.inc* file that is distributed with GWM-2000 and GWM-2005.) The six options now provided are

IRM = 0: An existing nonformatted file containing the response matrix is read and the LP solved.

IRM = 1: The response matrix is computed by GWM and saved to a nonformatted file. The LP is then solved.

IRM = 2: The response matrix is computed by GWM and the LP solved. The response matrix is not saved or printed.

IRM = 3: The response matrix is computed by GWM and printed to a formatted file. The LP is then solved.

IRM = 4: The response matrix is computed by GWM, saved to a nonformatted file, and printed to a formatted file. The LP is then solved.

IRM = 5: An existing nonformatted file containing the response matrix is read and the LP solved. The response matrix is printed to a formatted file.

If IRM is set to 0, 1, or 3, a filename needs to be specified by using variable RMNAME1 (item 4f) in the **SOLN** input file. If IRM is set to 4 or 5, two filenames need to be specified by using variables RMNAME1 and RMNAME2 in item 4f. The first filename (RMNAME1) is the nonformatted file and the second filename (RMNAME2) is the formatted file.

Expanded capabilities to write the response matrix also were added to the no solution (NS) option of GWM-2005 specified in the **SOLN** file. Two options are allowed by specifying values of 1 or 3 for variable IRM in item 2c of the **SOLN** file (if no value is specified for IRM, a default value of 1 is assumed by GWM):

IRM = 1: the response matrix is computed by GWM and saved to a nonformatted file, or

IRM = 3: the response matrix is computed by GWM and printed to a formatted file.

An example response matrix written by GWM to a formatted file is given in figure 3. The response coefficients shown in the figure are those for the DEWATER sample problem described in Ahlfeld and others (2005). The LP solution option is used in this sample problem, and variable IRM was set to 3 to generate the response coefficients shown in the figure. The response matrix is written row by row in a format similar to that used to write the MODFLOW head solution. Each row corresponds to one of the simulation-based constraints (in this case, head constraints at the 10 locations named b-01, b-02, and so forth). The rows are written in the order in which each of the head and streamflow constraints are read for a GWM run: first, the response coefficients corresponding to head-based constraints are listed in the order in which the head-based constraints were read from the **HEDCON** file; second, the response coefficients corresponding to streamflow-based constraints (if any) are listed in the order in which the streamflow-based constraints were read from the **STRMCON** file. Each column of the response matrix represents a flow-rate decision variable (in this case, withdrawal wells at the seven candidate well sites). The flow-rate decision variables are listed in the order in which they appear in the **DECVAR** file. With reference to the values shown in figure 3, the first column of values is the change in head (in feet) at each of the 10 constraint locations divided by the change in pumping rate (that is, the perturbation value specified in the GWM input files) at flow-rate decision variable Q1 (in cubic feet per day). The units of the response coefficients shown in figure 3 are therefore feet of drawdown per cubic foot per day of pumping. Note that pumping at well Q1 has the largest effect on heads at constraint locations b-02 and b-01, followed by constraints b-04, b-03, and so forth.

Figure 3. Example response-matrix output file for DEWATER sample problem.

```
GWM RESPONSE MATRIX
   ROWS IN READ-ORDER FOR SIMULATION-BASED CONSTRAINTS
      HEAD CONSTRAINTS IN ORDER READ FROM HEDCON
      STREAM CONSTRAINTS IN ORDER READ FROM STRMCON
   COLUMNS IN READ-ORDER FOR FLOW-RATE DECISIONS VARIABLES
   ------------------------------------------------------------


                  1              2              3              4              5
                  6              7
.............................................................................
      1    0.101949E-01   0.778914E-02   0.797034E-02   0.750478E-02   0.660235E-02
           0.517529E-02   0.463308E-02
      2    0.103054E-01   0.103421E-01   0.929592E-02   0.761145E-02   0.764697E-02
           0.579649E-02   0.494582E-02
      3    0.782590E-02   0.103054E-01   0.804271E-02   0.663787E-02   0.761145E-02
           0.599856E-02   0.491298E-02
      4    0.991186E-02   0.752346E-02   0.894489E-02   0.972287E-02   0.734398E-02
           0.579601E-02   0.534809E-02
      5    0.755933E-02   0.100196E-01   0.901600E-02   0.737916E-02   0.982850E-02
           0.747409E-02   0.582287E-02
      6    0.731580E-02   0.642287E-02   0.759578E-02   0.961431E-02   0.724018E-02
           0.636208E-02   0.631907E-02
      7    0.742033E-02   0.745515E-02   0.890618E-02   0.971859E-02   0.975333E-02
           0.832910E-02   0.734480E-02
      8    0.628342E-02   0.668314E-02   0.734480E-02   0.759432E-02   0.883312E-02
           0.110537E-01   0.890618E-02
      9    0.479245E-02   0.493955E-02   0.534692E-02   0.568859E-02   0.595585E-02
           0.742033E-02   0.114198E-01
     10    0.463179E-02   0.487710E-02   0.519641E-02   0.541579E-02   0.585875E-02
           0.758425E-02   0.100196E-01
```

Change to Solution Variable NPGNMX: Turning Off Reexecution of the GWF Process

GWM performs a series of tests on the outcome of each run of the MODFLOW GWF Process. These tests include checking for flow-process convergence, adequate precision of the response coefficients, and dewatering of cells that are included in head constraints. These tests are discussed in detail on pages 32 and 33 of Ahlfeld and others (2005), and are controlled by user-specified variables NSIGDIG, AFACT, PGFACT, and NPGNMX in the **SOLN** file. If these tests indicate that the GWF Process run was unsuccessful, then GWM will automatically rerun the flow-process simulation with adjusted input settings. Setting NPGNMX to a value greater than 0 indicates the number of attempts that will be made to achieve a successful flow-process run; setting NPGNMX to 0 eliminates testing of the flow-process results except to confirm that the flow process has converged. If NPGNMX is set to 0, and GWM determines that the GWF Process has failed to converge, then the GWM run will be terminated.

Writing of Flow-Rate Decision Variables in the Optimal Solution to a MODFLOW Well-Type File

A feature has been added to GWM to allow the value of the flow-rate decision variables in the optimal solution to be written to an output file called **GWMWFILE**. The flow-rate decision variables are written in a format that is similar to that of the input file for the MODFLOW Well (**WEL**) Package, including the layer, row, and column location of each flow-rate decision variable. Optimal flow rates are written for all flow-rate decision variables for each stress period, even if the flow rate for the well is zero. When a flow-rate decision variable is applied over multiple cells, each of the cells is identified along with the portion of optimal pumping for that cell. When a flow-rate decision variable is applied over multiple stress periods, the flow rate and cell location for that variable are written for each stress period during which it is active. For problems with no unmanaged wells, the completed **GWMWFILE** can be used as the basis for a **WEL** input file that would reproduce the optimal results; if unmanaged wells are present, then the user will need to merge the **GWMWFILE** with the unmanaged **WEL** input file to reproduce the optimal results.

Output to the **GWMWFILE** file is activated by providing a positive-integer value to variable GWMWFILE in the **DECVAR** file (see "Input Instructions and Output Files" section of the report). The value of GWMWFILE is the unit number of the file to which output will be written. The unit number should be associated with a file that is defined in the MODFLOW **NAME** file by the file type *DATA*. If the value specified for GWMWFILE is zero, negative, or blank, then the output will not be written. If SLPITPRT in the **SOLN** file is set to 2, then GWMWFILE will be written at the end of each SLP iteration with only the most recent solution appearing in the file.

An example of the use of **GWMWFILE** with GWM-2005 is provided for the SUPPLY2 sample problem first described in Ahlfeld and others (2005) and modified as described in the "Sample Problems" section of this report. First, the **DECVAR** file was modified to set variable GWMWFILE equal to 20:

```
#SUPPLY2 Sample Problem, DECVAR file
#August 14, 2006
1 20            #1-IPRN GWMWFILE
```

Next, the MODFLOW **NAME** file was modified to include a record for the file to which the flow rates will be written (file *supply2.gwmwell*):

```
DATA       20    supply2.gwmwell
```

The output file (*supply2.gwmwell*), which is reproduced in figure 4, contains the optimal flow rates for each managed well in the solution. Note that there are nonzero withdrawal rates for each of the 12 stress periods in this transient problem. The output file also lists the name of each flow-rate decision variable (from the **DECVAR** file) and the number of model cells over which the flow rate is distributed (variable NC in the **DECVAR** file).

Figure 4. Output listing for GWMWFILE for SUPPLY2 sample problem.

```
2        0                              Stress Period:    1
1       12      11 -0.3483296E+04      Flow Variable:Q1          NC=  1
1       16      17 -0.5000000E+05      Flow Variable:Q2a         NC=  1
3        0                              Stress Period:    2
1       12      11 -0.3483296E+04      Flow Variable:Q1          NC=  1
1       16      17 -0.2755393E+05      Flow Variable:Q2b         NC=  1
1       14      25 -0.4896278E+05      Flow Variable:Q4a         NC=  1
2        0                              Stress Period:    3
1       12      11 -0.3483296E+04      Flow Variable:Q1          NC=  1
1       16      17 -0.2690226E+05      Flow Variable:Q2c         NC=  1
3        0                              Stress Period:    4
1       12      11 -0.3483296E+04      Flow Variable:Q1          NC=  1
1       16      17 -0.2673937E+05      Flow Variable:Q2d         NC=  1
1       14      25 -0.8068633E+04      Flow Variable:Q4b         NC=  1
2        0                              Stress Period:    5
1       12      11 -0.3483296E+04      Flow Variable:Q1          NC=  1
1       16      17 -0.5000000E+05      Flow Variable:Q2a         NC=  1
3        0                              Stress Period:    6
1       12      11 -0.3483296E+04      Flow Variable:Q1          NC=  1
1       16      17 -0.2755393E+05      Flow Variable:Q2b         NC=  1
1       14      25 -0.4896278E+05      Flow Variable:Q4a         NC=  1
2        0                              Stress Period:    7
1       12      11 -0.3483296E+04      Flow Variable:Q1          NC=  1
1       16      17 -0.2690226E+05      Flow Variable:Q2c         NC=  1
3        0                              Stress Period:    8
1       12      11 -0.3483296E+04      Flow Variable:Q1          NC=  1
1       16      17 -0.2673937E+05      Flow Variable:Q2d         NC=  1
1       14      25 -0.8068633E+04      Flow Variable:Q4b         NC=  1
2        0                              Stress Period:    9
1       12      11 -0.3483296E+04      Flow Variable:Q1          NC=  1
1       16      17 -0.5000000E+05      Flow Variable:Q2a         NC=  1
3        0                              Stress Period:   10
1       12      11 -0.3483296E+04      Flow Variable:Q1          NC=  1
1       16      17 -0.2755393E+05      Flow Variable:Q2b         NC=  1
1       14      25 -0.4896278E+05      Flow Variable:Q4a         NC=  1
2        0                              Stress Period:   11
1       12      11 -0.3483296E+04      Flow Variable:Q1          NC=  1
1       16      17 -0.2690226E+05      Flow Variable:Q2c         NC=  1
3        0                              Stress Period:   12
1       12      11 -0.3483296E+04      Flow Variable:Q1          NC=  1
1       16      17 -0.2673937E+05      Flow Variable:Q2d         NC=  1
1       14      25 -0.8068633E+04      Flow Variable:Q4b         NC=  1
```

Use of the GWM OUT File in Place of the GLOBAL File

The only substantial change that was made to the structure of the GWM input and output files for GWM-2005 is the way in which output from a GWM run is saved to a file. This change was necessary because MODFLOW-2005 does not support the **GLOBAL** output file, which was an option in MODFLOW-2000. Instead, all output from a MODFLOW-2005 simulation is written to the MODFLOW **LIST** file. GWM-2000 used the **GLOBAL** file to save information about the GWM run, and the name of the **GLOBAL** file was specified in the MODFLOW **NAME** file. In order to retain a separate file for output that is specific to the GWM Process, GWM-2005 now writes GWM related information to the GWM **OUT** file. Instructions for specifying the name of the **OUT** file are provided in the "Input Instructions and Output Files" section of this report.

Specification of Streamflow Constraints with the Streamflow-Routing (SFR) Package

Values of streamflow and streamflow depletion at streamflow-constraint locations were calculated in previous versions of GWM by use of the original Streamflow-Routing Package of MODFLOW called STR (Prudic, 1989). GWM-2005 has been updated to allow the specification of streamflow constraints with either the original STR Package or the updated SFR Package (Prudic and others, 2004; Niswonger and Prudic, 2005). Typically, a modeler will use either the STR or SFR Packages to simulate streamflow for a particular problem. However, GWM does allow the use of STR and SFR Packages simultaneously to simulate stream-flow, but when streamflow constraints are used in a groundwater-management formulation, the user cannot have the STR and SFR Packages active on a grid simultaneously. For example, if a management problem includes streamflow constraints on the parent grid, the user can specify either the STR or SFR Packages to simulate streamflow on the parent grid, but not both packages. If this same problem includes simulation of streamflow on a child grid, but not the specification of streamflow constraints on that grid, then both the STR and SFR Packages could be used simultaneously on the child grid. Moreover, the user could specify stream-flow constraints on one grid with either the STR or SFR Packages while simultaneously specifying streamflow constraints with the other package on the second grid.

The SFR Package is similar in most aspects to the STR Package. For example, both packages use the continuity equation to route surface-water flow through one or more simulated rivers, streams, canals, or ditches. One of the most significant differences between the two packages is that in SFR, stream depth is calculated at the midpoint of each reach instead of the beginning of each reach, as was done in the original STR Package. This approach allows for the addition and subtraction of water from runoff, precipitation, and evapotranspiration within each stream reach. Because the SFR and STR Packages use different methods to calculate stream depth within each reach, the model-calculated flow rates across streambeds will likely differ between the SFR and STR Packages. Unlike the STR Package, the SFR Package also supports simulation of unsaturated flow beneath streams (Niswonger and Prudic, 2005).

As with the STR Package, the SFR Package is best suited for modeling long-term changes (months to hundreds of years) in groundwater flow by using averaged flows in streams; the packages are not recommended for modeling the transient exchange of water between streams and aquifers when the objective is to examine short-term (minutes to days) effects caused by rapidly changing streamflows (Prudic and others, 2004). An example of the use of the SFR Package in GWM-2005 is provided in the "Sample Problems" section of this report.

Option to Adjust the Closure Criteria for GWF Process Runs

GWM executes multiple runs of the GWF Process during its solution procedure. At the conclusion of each run, GWM determines whether the results of the GWF Process are acceptable. One criterion for acceptability is the convergence status of the solution obtained by the GWF Process at the end of each time step. Depending on the solver package used, the applicable GWF convergence criteria may be defined by HCLOSE and/or RCLOSE. For multiple-grid problems, both the head closure criterion (HCLOSELGR) and flux closure criterion (FCLOSELGR) are relevant. In some cases, the GWM-2005 user may want to consider a solution at a time step acceptable even if the applicable convergence criteria are not met. This might be the case, for example, if the convergence failure is a result of poor head convergence far from the part of the model domain in which management is being done.

GWM-2005 allows the user the option to adjust the criterion for accepting the results of a GWF Process run. This is done by specification of variable CRITMFC in the **SOLN** file. For a single grid (that is, simulations that do not use LGR), the options for CRITMFC are

CRITMFC = 0.0: convergence of the GWF Process will be assumed only if the applicable GWF convergence criteria (HCLOSE and/or RCLOSE) have been met at the end of each time step (this is the standard approach for the GWF Process of MODFLOW).

CRITMFC > 0.0: the results of the GWF Process will be accepted if the applicable GWF convergence criteria (HCLOSE and/or RCLOSE) have not been met, but the percent discrepancy between the inflow and outflow rates for each time step as reported in the GWF Process volumetric budget is less than or equal to the value specified for CRITMFC. For example, if CRITMFC is specified as 0.1, then GWF Process results will be accepted if the percent discrepancy between the inflow and outflow rates of the volumetric budget at the end of each time step is less than or equal to 0.1 percent.

CRITMFC < 0.0: the results of the GWF Process will be accepted regardless of the GWF convergence criteria (HCLOSE and/or RCLOSE) or the percent discrepancy.

For multiple grids (that is, simulations with LGR), the CRITMFC options are

CRITMFC = 0.0: convergence of the GWF Process will be assumed only if the HCLOSELGR and FCLOSELGR criteria have been met at the end of each time step (this is the standard approach for an LGR simulation).

CRITMFC > 0.0: the results of the GWF Process will be accepted if the HCLOSELGR or FCLOSELGR criteria have not been met, but the percent discrepancy between the inflow and outflow rates for each time step as reported in the GWF Process volumetric budget for all grids is less than or equal to the value specified for CRITMFC. For example, if CRITMFC is specified as 0.1, then GWF Process results will be accepted if the percent discrepancy between the inflow and outflow rates of the volumetric budget for all grids at the end of each time step is less than or equal to 0.1 percent.

CRITMFC < 0.0: the results of the GWF Process will be accepted regardless of the HCLOSELGR or FCLOSELGR criteria or the percent discrepancy.

Users are cautioned that when CRITMFC is set to a value other than 0.0, simulated heads (and, optionally, fluxes across parent-child interfaces) may not be as accurate as would normally be desired for calculation of response coefficients, and that the accuracy of the model is reflected only by the accuracy of the water-budget calculations as defined by the CRITMFC closure criterion. The user should review the heads and other state variables (such as flow rates to simulated streams) that are calculated for the final (optimal) flow-process simulation to be sure they are reasonable.

GWM-2005 with Local Grid Refinement

This section outlines the integration of the GWM Process with the LGR capability of MODFLOW-2005. In the discussion that follows, it is important to keep in mind that although information for a particular groundwater-management formulation must be provided to GWM for each of the parent and child models that are included in the simulation, GWM assembles this information into a single optimization formulation that is then solved over the parent and all child models.

Formulation of Groundwater-Management Problems with Both Parent and Child Models

The formulation process for a GWM problem with multiple models is substantially the same as for a problem with only a single model—specifically, the user must define a set of decision variables, an objective function, and a set of constraints. When multiple models are used, decision variables and constraints may be placed on any of the model grids with some restrictions, whereas only a single objective function is used. The paragraphs below describe the capabilities for the formulation of a management problem that extends over multiple models.

Decision Variables

Flow-rate decision variables can be placed on the parent grid and (or) any of the child grids. A **DECVAR** input file must be defined for each model that includes flow-rate decision variables. Each flow-rate decision variable is assigned a name that must be unique over all models; that is, the same decision variable name cannot be used in more than one model. Flow-rate decision variables may include multiple cells, but all of these cells must be in the same grid. Because the shared nodes between the parent and child grids are used as boundary conditions in the LGR solution process, flow-rate decision variables should not be placed in the cells in which shared nodes are located.

External variables are not grid specific and therefore can be defined in the **DECVAR** file(s) for any of the parent or child models. Binary variables are also not grid specific and can be defined in the **DECVAR** file(s) for any of the parent or child models. Binary variables can be associated with flow-rate decision variables on more than one grid by using the unique names given to each flow-rate decision variable in each of the

DECVAR files. For example, a binary variable might be associated with two flow-rate decision variables, one defined for the parent grid and one for a child grid, and be defined in the **DECVAR** file of the child model. An example of the use of binary variables that are associated with flow-rate decision variables defined for multiple models is provided in the SUPPLY-3GRID sample problem later in the report.

Each model that includes a **DECVAR** file with flow-rate and (or) external decision variables must also include a **VARCON** file in which the upper and lower bounds on the variables are defined (see Constraints section below).

Objective Function

A single objective function is required for the management problem and must be specified in the **OBJFNC** file of the parent model. The objective function can include any or all of the flow-rate, external, and (or) binary decision variables defined in the **DECVAR** files for any of the parent and (or) child models and uses the unique names given to decision variables in any of the **DECVAR** files.

Constraints

A **VARCON** file that specifies the upper and lower bounds on flow-rate and external decision variables must be created for any model for which flow-rate and external decision variables have been specified. Moreover, an input record must be entered in the **VARCON** file for each of the flow-rate and external variables defined in the corresponding **DECVAR** file.

GWM-2005 allows the use of linear-summation constraints that include decision variables from more than one model. For example, a constraint might be specified that requires that the sum of withdrawal rates from several wells in both parent and child models exceed a specified minimum rate. The **SUMCON** file that is used to specify the linear-summation constraints for a problem must be included in the parent model. The user specifies the unique names for the decision variables that have been specified in each of the **DECVAR** files for the problem. An example of the use of linear-summation constraints with decision variables defined for multiple grids is provided in the SEAWATER-LGR sample problem later in the report.

Head-based and streamflow constraints can be specified for any model, and these constraint types are specified in **HEDCON** or **STRMCON** files for the model grids to which the constraints are applied. Head-difference and gradient constraints cannot be specified for cells on different grids; that is, both cells that are specified for a head-difference or gradient constraint must be on the same grid. Also, because the shared-node method of the LGR approach is based on the use of specified-head and specified-flux boundary conditions for the cells in which the shared nodes are located, head and streamflow constraints should not be applied to these shared-node cells.

Integration of LGR into the GWM Process

A simplified flowchart that shows the integration of the Groundwater Flow Process with the LGR capability and GWM Process is given in figure 5. As in earlier implementations of the GWM Process, the GWM algorithm wraps two loops around the GWF Process computations. The outer nonlinear iteration loop provides a means for iteration of the LP solution approach for nonlinear management problems. The inner GWM flow-process (perturbation) loop repeatedly executes the GWF Process for calculation of the response coefficients of the management problem. The LGR calculations are done as part of the GWF Process calculations. At the level of the GWM perturbation loop, the GWF Process can still be viewed as calculating the state of the system in response to imposed stresses, but now for multiple models; insertion of the LGR calculations into the GWF Process requires no fundamental change in the GWM solution process.

A single response matrix is assembled by GWM that consists of the responses for all decision-variable/state-variable pairs distributed across all models. Response coefficients are calculated by perturbing each flow-rate decision variable individually and then recording the system-state response (such as head or streamflow) to the perturbation of each constraint location on all of the grids. That is, the system-state response in any grid to perturbation at each flow-rate decision variable in any grid is determined, and the full response matrix is then constructed. The response matrix is then combined with other components of the formulation (the objective function, other constraints, and so forth) from all models to form a single complete optimization problem.

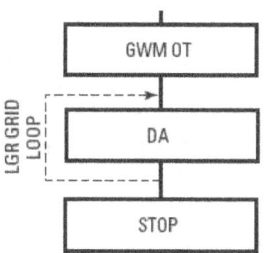

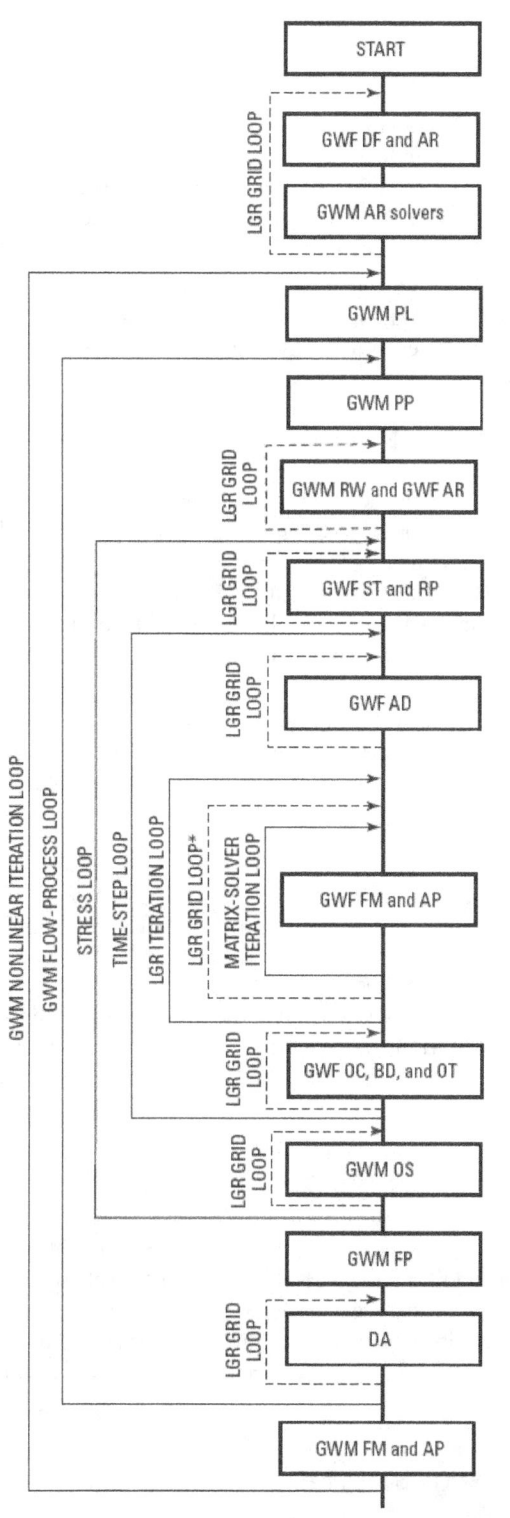

GWF and GWM Procedures:
DF Define
AR Allocate and Read
ST Stress
RP Read and Prepare
AD Advance a time step
FM Formulate equations
AP Solve equations
OC Output control
BD Calculate volumetric budget
OT Write output
DA Deallocate

GWM-Specific Procedures:
PL Prepare Loop
PP Prepare Process
RW Rewind
OS Observe State
FP Formulate Process

LGR Grid Loop: Executed once for each grid involved in LGR.

*This loop includes an LGR Cage/Shell loop for child grids.

(Flowchart continues from previous column)

(Flowchart continues on next column)

Figure 5. Simplified flowchart showing Groundwater Flow (GWF) Process with Local Grid Refinement (LGR) capability and Groundwater Management (GWM) Process.

GWM Solution Algorithms for Management Problems with Both Parent and Child Models

A single solution and output-control parameters (**SOLN**) file is specified for a management problem with multiple models and is included with the parent-model input files. Linear, nonlinear, and mixed-binary formulations can be solved when the LGR capability is used. Because GWM assembles a single optimization problem from information provided for the multiple models, there are no modifications required to the **SOLN** file for multiple models.

The link between the GWM Process and LGR capability results in multiple levels of iteration, as illustrated in figure 5 and in the introductory overviews of the GWM and LGR codes. Beginning with the GWF Process matrix solvers and then extending outward to iteration levels at the LGR and then GWM steps, these levels of iteration include the following: For most of the matrix solvers that are used for solution of the GWF Process (specifically, SIP, PCG2, and GMG), iteration is done for each model at each time step. LGR subsequently iterates between models to ensure agreement in heads and fluxes at shared nodes of the parent and child grids. GWM then makes repeated calls to the GWF Process for different values (perturbations) of the flow-rate decision variables, as well as making repeated iterations of the LP algorithm when solving a nonlinear problem with SLP. It is therefore imperative that the convergence criteria that are selected at each level of iteration be consistent with the criteria applied at other levels of iteration. Generally, this means that convergence criteria should be most strict (for example, HCLOSE values smallest) for the inner iterations and become slightly less restrictive for each higher level iteration. An example of the selection of closure (convergence) criteria for the various levels of iteration is provided in the SEAWATER-LGR sample problem later in this report.

When simulating nonlinear management problems with the SLP algorithm, the user must specify two convergence parameters, one for the value of the objective function (SLPZCRIT) and one for the values of the flow-rate decision variables (SLPVCRIT). These convergence criteria are described in detail in Ahlfeld and others (2005, p. 25–26).

GWM Output for Multiple Models

A full discussion of the output files that are produced by GWM-2005 is provided in the "Input Instructions and Output Files" section of this report. Only two considerations regarding the output files are relevant to the use of GWM with multiple models. First, a single GWM-2005 **OUT** file is generated for multiple models, and the name of the **OUT** file is specified in the **GWM** file of the parent model. Second, the names of the input files from which the decision variables and head and streamflow constraints are read for each model are written to the **OUT** file.

Input Instructions and Output Files

This section describes the input instructions and output files for GWM-2005 with LGR capability. Before solving an optimization problem with GWM-2005, the user must already have developed a groundwater-flow model (or both parent and child models if LGR capability is being used) of the study area based on the MODFLOW-2005 GWF Process. In its current form, GWM-2005 can only be used with the GWF and Observation (OBS) Processes of MODFLOW-2005.

GWM-2005 can be used in one of four modes:

- MODFLOW-2005 GWF Process without GWM or LGR capabilities,

- MODFLOW-2005 GWF Process with LGR capability but without GWM,

- GWM with the MODFLOW-2005 GWF Process but without LGR, and

- GWM with the MODFLOW-2005 GWF Process with the LGR capability.

Selection of one of the four modes is made on the basis of the type and contents of the file that is entered by the user in response to an onscreen prompt when GWM-2005 is first executed. Two types of files can be entered—a MODFLOW **NAME** file or an LGR **CONTROL** file, each of which is described in detail below. Implementation of each of the four modes is done as follows:

- A MODFLOW-2005 GWF-Process Simulation without GWM or LGR capabilities. GWM-2005 can be used to run the MODFLOW-2005 GWF Process without either the GWM Process or LGR capability active. In this case, the user enters the name of a MODFLOW **NAME** file. The **NAME** file contains the names of input and output files that are used by the MODFLOW GWF and OBS Processes only. The GWM-2005 Process is not activated.

- A MODFLOW GWF-Process Simulation with LGR capability but without GWM. In this case, a MODFLOW-2005 GWF Process simulation is done with multiple models but without groundwater management. At the GWM-2005 prompt, the user enters the name of an LGR **CONTROL** file. The LGR **CONTROL** file is distinguished from a **NAME** file by the presence of the keyword *LGR* as the first noncommented input. LGR reads its input data for the parent and child models from this control file. **NAME** files are specified for each of the parent and child models as part of the LGR **CONTROL** file. The GWM-2005 Process is not activated.

- A GWM Simulation without LGR. In this case, GWM-2005 is used to simulate a groundwater-management problem that consists of a single model covering the model domain. At the GWM-2005 prompt, the user enters the name of a MODFLOW **NAME** file. The **NAME** file contains the names of input and output files that are used by the MODFLOW GWF and OBS Processes as well as the name of the **GWM** file (see input instructions below).

- A GWM Simulation with the LGR capability. In this case, GWM-2005 is used to simulate a ground-water-management problem that consists of multiple models—a parent model and at least one child model. At the GWM-2005 prompt, the user enters the name of an LGR **CONTROL** file. **NAME** files are specified for each of the parent and child models as part of the LGR **CONTROL** file; the name of the **GWM** file is specified in the **NAME** file for the parent model and for any child model that includes components of the management problem.

MODFLOW NAME File

The MODFLOW **NAME** file contains the names of most input and output files used by MODFLOW and determines which MODFLOW program options are activated. The **NAME** file is read on unit 99. When LGR is used, a **NAME** file must be specified for each model in the LGR **CONTROL** file. Each **NAME** file contains one record (line) of information for each input and output file. Each record consists of as many as four variables, which are read in free format; the length of each record must be 299 characters or less. Comment records can be placed anywhere in the **NAME** file and are indicated by the # character in column one. Any text characters can follow the # character. Comment records have no effect on the run; their purpose is to allow users to provide documentation about a particular run.

Each record has the following format:

Ftype   Nunit   Fname   [Fstatus]

The variables are defined as follows:

Ftype—is the file type. Ftype may be entered in all uppercase, all lowercase, or any combination thereof.

Nunit—is the Fortran unit to be used when reading from or writing to the file. Any legal unit number on the computer being used can be specified except 99. Also, the unit number for the file must be unique; that is, it cannot be equal to any of the unit numbers used for other files specified in the **NAME** file.

Fname—is the name of the file, which is a character value. Pathnames may be specified as part of Fname.

Fstatus—is the optional file status, which applies only to file types **DATA** and **DATA(BINARY)** (see Harbaugh, 2005). Two values are allowed: OLD and REPLACE. "OLD" indicates that the file should already exist. "REPLACE" indicates that if the file already exists, then it should be deleted before a new file is opened. The default actions are to open the existing file if the file exists or create a new file if the file does not exist.

The listing file (Ftype = *LIST*) must be present in each **NAME** file and must be the first file listed in the **NAME** file. If a management model is to be solved, the **NAME** file must include a record that specifies Ftype = *GWM* to indicate that the GWM-2005 Process is active. The **GWM** file identified in this record contains information needed for the GWM Process. For a GWM-2005 run that solves a management model and uses LGR, a **GWM** file record must appear in the **NAME** file for the parent model; the **GWM** file record may optionally be included in the **NAME** file for the child models. When present, the *GWM* record must appear after the *LIST* record. Example input records for the **LIST** and **GWM** file types are

| *LIST* | 10 | list.gwm |
| *GWM* | 55 | input.gwm |

MODFLOW DISCRETIZATION File

When the LGR capability of GWM-2005 is used, the number of stress periods in a simulation (variable NPER), the length of each stress period (variable PERLEN), the number of time steps in each stress period (variable NSTP), and the time-step multiplier for each stress period (variable TSMULT) are the same for all models. These variables must be specified in the **DISCRETIZATION** (**DIS**) input file for each model, but GWM-2005 uses only the values specified in the **DIS** file of the parent model and ignores the values specified in the child model(s).

Considerations for Preparing MODFLOW-2005 Files for Use with LGR

A few issues must be considered when preparing MODFLOW-2005 input files for a simulation that uses LGR (the user is directed to the LGR documentation reports by Mehl and Hill, 2005 and 2007, for a complete description of the LGR capability):

- The parent and child models each require a MODFLOW **NAME** file and associated set of input files. The unit numbers defined in each of these MODFLOW files also need to be unique; a unit number used in the parent-model input and output files cannot be used for any of the child models.

- Because the parent and each of the child models are separate models, different MODFLOW packages can be used for each model.

- In the MODFLOW Basic Package input file for each child model, set IBOUND = IBFLG. Variable IBOUND is the boundary variable specified for every model cell (Harbaugh, 2005, p. 8–14). Variable IBFLG is set in the LGR **CONTROL** file and is a negative integer used to define the interface of the child grid with the parent grid. A unique value of IBFLG must be used for each child grid.

- In the LGR shared-node method, data input may require modification for the cells that form the interface between the parent-child grids if the Drain (DRN), Evapotranspiration (EVT), General-Head Boundary (GHB), Recharge (RCH), River (RIV), or Well (WEL) Packages are used at these cells. Except for the WEL Package, the flux calculated by these packages depends on the full cell area, either directly or through a user-defined conductance term. Cells at the interface of the parent model are truncated, but in the packages listed the cell areas are not modified. Thus, the user needs to adjust inputs so that the proper influence of the stress is represented in the model (see Mehl and Hill, 2005, p. 5–7).

- If the DE4 solver (Harbaugh, 1995) is used for one or more of the child models, the decomposition cannot be reused from a previous time step or internal iteration, even if the model is linear. Thus, IFREQ = 3 should be used.

LGR CONTROL File

At the beginning of a GWM-2005 run, the program prompts the user for either a MODFLOW **NAME** file or an LGR **CONTROL** file. If a MODFLOW **NAME** file is specified, GWM-2005 will simulate a single model; if an LGR **CONTROL** file is specified, GWM-2005 will simulate multiple models, each with its own MODFLOW **NAME** file. The LGR **CONTROL** file is distinguished from a MODFLOW **NAME** file by the presence of the keyword *LGR* as the first noncommented input variable. LGR reads its input data from this **CONTROL** file. Input for LGR is defined by 15 items, each of which is read in free format:

FOR EACH SIMULATION

0.   [#Text]

Item 0 is optional—# must appear in column 1. Item 0 can be repeated multiple times.

1.   ***LGR***

2.   NGRIDS

FOR THE PARENT MODEL (input items for the parent model must be listed before those for the child model or models)

3.   NAME FILE

4.   GRIDSTATUS

5.   IUPBHSV   IUPBFSV

FOR EACH CHILD MODEL (items 6 through 15 are repeated for each child model; the total number of repetitions is NGRIDS − 1)

6.   NAME FILE

7.   GRIDSTATUS

8.   ISHFLG   IBFLG   IUCBHSV   IUCBFSV

9.   MXLGRITER   IOUTLGR

10.   RELAXH   RELAXF

11.   HCLOSELGR   FCLOSELGR

12.   NPLBEG   NPRBEG   NPCBEG

13.   NPLEND   NPREND   NPCEND

14.   NCPP

15.   NCPPL (repeat NCPPL a total of NPLEND + 1 − NPLBEG times)

The variables are defined as follows:

LGR—is a keyword that must be entered in the first item to indicate to GWM-2005 that the LGR capability is used.

NGRIDS—is the total number of grids used in this simulation, including the parent grid and all of the child grids.

NAME FILE—is the name of the **NAME** file for either the parent or child model. The name can include the file path and is limited to 200 characters.

GRIDSTATUS—is a character variable and indicates whether the file listed in NAME FILE corresponds to a parent or child model. Two options are allowed:

　　PARENTONLY—the **NAME** file for the parent model; and
　　CHILDONLY—the **NAME** file for a child model.

IUPBHSV—is an integer value greater than zero that corresponds to the unit number for the file to which boundary heads will be saved for later use by the Boundary Flow and Head (BFH) Package for independent simulations. A file with this unit number needs to be opened in the **NAME** file of the parent model. A value of zero indicates that the file is not written. For the parent model, these are the complementary boundary conditions (see Appendix 2 of Mehl and Hill, 2005).

IUPBFSV— is an integer value greater than zero that corresponds to the unit number for the file to which boundary fluxes will be saved for later use by the BFH Package for independent simulations. A file with this unit number needs to be opened in the **NAME** file of the parent model. A value of zero indicates that the file is not written. For the parent model, these are the coupling boundary conditions (see Appendix 2 of Mehl and Hill, 2005).

ISHFLG—is a flag indicating whether heads from the parent-model simulation should be used as the starting heads for the child-model simulation(s). These heads apply to the interior of the child grid(s), not the boundary heads; for interior child-grid cells that overlap the parent grid, the heads of the corresponding parent cell are used. Two options are allowed:

ISHFLG = 1. Use heads from the parent-model simulation as the starting heads for the child model. No interpolation is applied. For steady-state simulations, this procedure can provide a good initial estimate of the head distribution and thus reduce computational time. For transient simulations, the procedure over-writes the initial condition of the child model defined in STRT of the Basic Package input file (Harbaugh, 2005) and therefore is not recommended.

ISHFLG = 0. Use the heads defined in STRT of the Basic Package for the child model.

IBFLG—is a negative integer used to define the interface of the child grid with the parent grid. Use this value around the perimeter of the child model IBOUND array. Do not use IBFLG or –IBFLG anywhere else in the parent or child IBOUND arrays. Use a unique value for each child grid.

IUCBHSV—is an integer value greater than zero that corresponds to the unit number for the file to which boundary heads will be saved for later use by the BFH Package for independent simulations. A file with this unit number needs to be opened in the **NAME** file of the child model. A value of zero indicates that the file is not written. For the child model, these are the coupling boundary conditions (see Appendix 2 of Mehl and Hill, 2005).

IUCBFSV— is an integer value greater than zero that corresponds to the unit number for the file to which boundary fluxes will be saved for later use by the BFH Package for independent simulations. A file with this unit number needs to be opened in the **NAME** file of the child model. A value of zero indicates that the file is not written. For the child model, these are the complementary boundary conditions (see Appendix 2 of Mehl and Hill, 2005).

MXLGRITER—is the maximum number of LGR iterations. Twenty iterations are sufficient for most problems. See "Closure Criteria for LGR Iterations" section of the report by Mehl and Hill (2005). Set MXLGRITER to 1 for a one-way coupling (see Mehl and Hill, 2005).

IOUTLGR—is an integer value that is a flag that controls printing of the maximum head and flux change for each LGR iteration. For the maximum head change, the head value and corresponding layer, row, and column of the child model is listed. For the maximum flux change, the flux value and corresponding layer, row, and column of the parent model is listed. If IOUTLGR < 0, output is written to the screen; if IOUTLGR > 0, output is written to the child listing file; and if IOUTLGR = 0, no results are written.

RELAXH, RELAXF—are real numbers equal to the relaxation factors for heads and fluxes, respectively. Values of RELAXH and RELAXF less than 1.0 and greater than zero are needed for convergence of the LGR iterations. Typically, values of about 0.5 produce convergent solutions. Values less than 0.5 may be needed when the LGR iterations have difficulty converging. In cases in which the LGR iterations converge, values greater than 0.5 may reduce the number of iterations needed for convergence. Convergence problems can be diagnosed by printing the maximum head and flux changes (IOUTLGR ≠ 0) to determine if the head and flux changes are decreasing (converging) or increasing (diverging) as the LGR iterations proceed.

HCLOSELGR—is a real number equal to the head-closure criterion for the LGR iterations. The closure criterion is based on heads for the child-interface nodes. This closure criterion is satisfied when the maximum absolute head change between successive LGR iterations is less than HCLOSELGR (see equation 8b in Mehl and Hill, 2005).

FCLOSELGR—is a real number equal to the flux-closure criterion for the LGR iterations. This closure criterion is based on fluxes into the parent-interface nodes. This closure criterion is satisfied when the maximum absolute relative flux change between successive LGR iterations is less than FCLOSELGR (see equation 8a in Mehl and Hill, 2005).

NPLBEG—is the layer number of the parent grid in which refinement (that is, the child model) begins. Refinement must begin in the top layer of the model, so NPLBEG must equal 1.

NPRBEG—is the row number of the parent grid in which refinement begins (cannot equal 1).

NPCBEG—is the column number of the parent grid in which refinement begins (cannot equal 1).

NPLEND—is the layer number of the parent grid in which refinement ends. NPLEND must be greater than or equal to NPLBEG.

NPREND—is the row number of the parent grid in which refinement ends. NPREND must be greater than NPRBEG, and NPREND cannot equal the number of rows in the parent grid.

NPCEND—is the column number of the parent grid in which refinement ends. NPCEND must be greater than NPCBEG, and NPCEND cannot equal the number of columns in the parent grid.

NCPP—is the number of child cells that span the width of a single parent cell along rows and columns. This must be an odd integer greater than 1, and is applied to both rows and columns.

NCPPL—is the number of child cells that span the depth of each parent layer. Each value must be an odd integer greater than or equal to 1. One value is read for each refined parent layer. The number of values needs to equal (NPLEND + 1 – NPLBEG). Values can be 1 (which results in no vertical refinement for the layer) only in layers above the bottom of the child model, unless refinement extends all the way to the bottom of the parent model. For refinement that does not extend to the bottom of the parent model, the refinement terminates at the shared node. For example, for the simulation condition shown in figure 2b, the values 5 3 would be entered for NCPPL.

GWM Process Files

Input files for the GWM Process consist of the **GWM** file and several supporting files. The GWM Process is activated by placing **GWM** file type in the MODFLOW **NAME** file along with the name of the **GWM** file. The **GWM** file, in turn, contains keywords and filenames for the components of the management problem. When the management problem is solved for multiple models, an LGR **CONTROL** file is required. It will include the **NAME** files for each model. The **NAME** files may each contain **GWM** files, with each **GWM** file containing references to the files that contain components of the management problem for the corresponding model.

Four types of information about the management problem are specified in the input files referenced in the **GWM** file: the decision variables, objective function, and constraints of the management problem and the solution and output-control parameters. GWM-2005 reserves Fortran unit 99 for input purposes. In contrast to the MODFLOW **NAME** file, unit numbers are not required for the files named in the **GWM** file. Instead, the program automatically determines available unit numbers and uses them for input and output.

For the most part, the general structure of the input formats for the GWM Process files are consistent with those for other MODFLOW processes; users of GWM should review the input instructions for MODFLOW given in Harbaugh (2005). Input for each GWM Process file is grouped by numbered items, and each item consists of input variables. The first item in each of the input files is Item 0, which can be used for comment lines but is optional. Some items consist of several variables, and the item can be repeated multiple times. The input data for each item must start on a new record. Each record is limited to a length of 199 characters. An input variable may include a single value or multiple values. Variables are defined after all the items are listed.

Each input variable has a data type, which can be Real, Integer, or Character. Integers are whole numbers and must not include a decimal point or exponent. Real numbers can include a decimal point and an exponent; if no decimal point is included in the entered value, then the decimal point is assumed to be at the right side of the value. Any printable character is allowed for character variables. Unlike the GWF Process, variables used by GWM that start with the letters I-N are not necessarily integers and those that start with the letters A-H and O-Z are not necessarily real numbers. Data types are specified for each input variable.

Free formatting is used for GWM input. With free formatting, values are not required to occupy a fixed number of columns in a record. Each value can occupy one or more columns as required to represent it; however, the values must still be included in the prescribed order. One or more spaces or a single comma optionally combined with spaces must separate adjacent values. Also, a numeric value of zero must be explicitly represented by 0 and not by one or more spaces.

Units of values used in the GWM Process should be consistent with the units used in the other MODFLOW data-input files.

GWM File

The **GWM** file is formatted in a manner similar to the MODFLOW **NAME** file. A series of records are read that have the following format:

Ftype   Fname

Ftype is one of several keywords, and Fname is a path name of the relevant computer file. Except for keyword *OUT*, each of the keywords triggers the reading of a file that will be referred to with the same name as the keyword. Unit numbers are determined internally by GWM and do not need to be specified in the **GWM** file. The entire record including the Fname entry is limited to 199 characters in length. Keywords can be specified in either uppercase or lowercase letters. Comment lines may appear anywhere in the **GWM** file and are indicated by the # character in the first column of the record. Keywords may appear in any order except for the *OUT* keyword, which must be the first keyword in the file if it is used. The keywords that are suitable for inclusion in a **GWM** file depend on the type of problem. If the problem is a single model, then only a single **GWM** file is provided. If the problem is multimodel, then a **GWM** file is required for the parent model and may be provided for child models. The following keywords are available in GWM-2005:

OUT—a filename for all output from the GWM Process may be assigned here. If the *OUT* keyword is not specified, a default name of "GWM.OUT" will be used, and the output file will be written to the directory in which program execution occurs. The *OUT* keyword is not allowed if the **GWM** file is for a child model.

DECVAR—the Fname associated with this keyword identifies the **DECVAR** file that provides information about the decision variables. For single-model problems, the *DECVAR* keyword is required. For multimodel problems, the **DECVAR** file is provided for every model that includes decision variables. The **GWM** file for at least one model, although not necessarily the parent model, must contain a *DECVAR* keyword.

VARCON—the **VARCON** file defined in this record provides information on the lower and upper bounds specified for the decision variables defined in the **DECVAR** file. If the *DECVAR* keyword appears in a **GWM** file, then the *VARCON* keyword must also appear.

OBJFNC—the **OBJFNC** file provides information about the objective function. The *OBJFNC* keyword must appear in the **GWM** file for single-model problems and in the parent model **GWM** file for multimodel problems. The *OBJFNC* keyword is not allowed in the **GWM** files of child models of multimodel problems.

SUMCON, HEDCON, and *STRMCON*—the **GWM** file may include up to three additional files that provide information about summation constraints, head constraints, and streamflow constraints that are allowed in GWM. None of these keywords are required in a **GWM** file. For multimodel problems, the *SUMCON* keyword can appear only in the parent model, whereas the *HEDCON* and *STRMCON* keywords may appear in the **GWM** files for parent or child models.

SOLN—the **SOLN** file provides information about the solution and output-control parameters. The *SOLN* keyword must appear in the **GWM** file for single-model problems and in the parent model **GWM** file for multimodel problems. The *SOLN* keyword is ignored in the **GWM** files of child models of multimodel problems.

The requirements for GWM files and keywords are summarized in table 1 and as follows. When the LGR capability is not active, each GWM run requires specification of a single GWM file. The GWM file must contain the **DECVAR, OBJFNC, VARCON,** and **SOLN** file types; the **OUT, SUMCON, HEDCON,** and **STRMCON** file types are optional. When the LGR capability is active, a **GWM** file must be in the **NAME**

Table 1. GWM file requirements for simulations with and without LGR.

File type	Simulation without LGR	Simulation with LGR	
		Parent model	Child model(s)
GWM	Required	Required	Optional
DECVAR	Required	Optional[1]	Optional[1]
OBJFNC	Required	Required	None
VARCON	Required	Optional[2]	Optional[2]
SUMCON	Optional	Optional[3]	None
HEDCON	Optional	Optional	Optional
STRMCON	Optional	Optional	Optional
SOLN	Required	Required	None
OUT	Optional	Optional	None

[1]At least one **DECVAR** file and associated **VARCON** file must be specified in either the parent or child models or both.

[2]A **VARCON** file must be specified if a **DECVAR** file is specified for the model.

[3]Constraints specified in a **SUMCON** file in the parent model may reference decision variables defined on the parent grid or any of the child grids.

file for the parent model. The parent **GWM** file specifies files that are universal to solving the GWM problem, namely, the **OBJFNC** and **SOLN** files. The parent-model **GWM** file also may optionally include **OUT**, **DECVAR**, **VARCON**, **SUMCON**, **HEDCON**, and **STRMCON** file types (a **VARCON** file must be specified if a **DECVAR** file is specified). The **GWM** file should be specified for a child model only if the child model contains decision variables, head constraints, or streamflow constraints. The **GWM** file for each child model may include **DECVAR**, **VARCON**, **HEDCON**, and **STRMCON** file types (a **VARCON** file must be specified if a **DECVAR** file is specified); file types **OUT**, **OBJFNC**, **SUMCON**, and **SOLN** may not be specified in a child **GWM** file. At least one **GWM** file, in either the parent or child model must contain a **DECVAR** file and associated **VARCON** file. If summation constraints are included in the problem, the **SUMCON** file must be listed in the **GWM** file for the parent model, although constraints may reference decision variables defined on any grid.

Decision Variable (**DECVAR**) File

This file is used to define the three types of decision variables that may be defined in a management model. The primary decision variables are the flow rates (either withdrawal or injection) at each managed well site. Flow-rate decision variables must be defined in the **DECVAR** file that is associated with the grid in which the flow is simulated. A single well site may have more than one flow-rate decision variable associated with it. Moreover, a single flow-rate decision variable can extend over one or more model cells and can be active during one or more stress periods. The cells that make up each flow-rate decision variable, however, must be completely within either the parent grid or one of the child grids; the cells that contain flow-rate decision variables cannot cross grid interfaces. The second type of decision variable is an external variable. External variables do not have a direct effect on the system state variables and are not assigned to a specific location in the model; for this reason, they can be defined in the **DECVAR** file(s) for any of the parent or child models. The third type of decision variable is a binary variable used to define the status of each flow-rate or external decision variable as active (for example, the site is constructed) or inactive (the site is not constructed). Binary variables have a value of 0 (inactive) or 1 (active). One or more flow-rate and external decision variables are associated with each binary variable. These associated variables may come from any of the flow-rate or external decision variables defined in the parent and child **DECVAR** files.

The **DECVAR** file includes five input items:

0.   [#Text]

Item 0 is optional—# must appear in column 1. Item 0 can be repeated multiple times.

1.   IPRN   GWMWFILE

2.   NFVAR   NEVAR   NBVAR

3a.   The following records are read for each of the NFVAR flow-rate decision variables:
      FVNAME   NC   LAY   ROW   COL   FTYPE   FSTAT   WSP

3b.   If NC>1 in record 3a, then the following record is read NC times, and the values of LAY, ROW, and COL read in record 3a are ignored:
      RATIO   LAY   ROW   COL

4.   The following record is read for each of NEVAR external decision variables:
      EVNAME   ETYPE   ESP

5.   The following record is read for each of NBVAR binary decision variables:
      BVNAME   NDV   BVLIST

The variables are defined as follows:

Text—is a character variable up to 199 characters long that starts in column 2. Any characters can be included in Text. Lines beginning with # are restricted to the first lines of the file. Text is printed when the file is read.

IPRN—is an integer variable that describes the amount of output that is written to the GWM **OUT** file. IPRN must be specified as either 0 or 1. When IPRN equals 0, a minimum amount of information about the decision variables is written to the GWM output file; when IPRN equals 1, detailed information about the decision variables is written to the GWM output file.

GWMWFILE—is an integer variable equal to the unit number of the file to which the values of the flow-rate decision variables at the optimal solution will be written. If GWMWFILE is set to zero, no output will be written. The flow rates are written in a format similar to the format used in the MODFLOW well (**WEL**) file. The unit number should be associated with a ***DATA*** file type defined in the MODFLOW **NAME** file. (See section "Writing of Flow-Rate Decision Variables in the Optimal Solution to a MODFLOW Well-Type File" for more details.)

NFVAR—is an integer variable equal to the number of flow-rate decision variables. NFVAR must be greater than 0. Only one flow-rate decision variable can be defined for a particular well site and set of stress periods with the exceptions that both a withdrawal variable (FTYPE=W) and an injection variable (FTYPE=I) can be defined for the site.

NEVAR—is an integer variable equal to the number of external decision variables. NEVAR must be greater than or equal to 0.

NBVAR—is an integer variable equal to the number of binary variables. NBVAR must be greater than or equal to 0. If NBVAR is 0, binary variables are not included in the management formulation.

FVNAME—is a character variable up to 10 characters long that is a unique name designated for the flow-rate decision variable. Each name must be unique (that is, the same name cannot be used for more than one variable, or in more than one model). No spaces are allowed in the name. The end of the name is designated by a blank space.

NC— is an integer variable equal to the number of model cells over which the flow rate for decision-variable FVNAME is distributed. NC must be greater than or equal to 1. If NC equals 1, then all of the water withdrawn or injected at decision variable FVNAME is applied at the single model cell LAY, ROW, COL. If NC is greater than 1, then the flow rate calculated for decision variable FVNAME is distributed over the NC cells specified in record 3b.

LAY, ROW, and COL—are integer variables equal to the layer, row, and column number of the model cell to which flow for decision-variable FVNAME will be assigned.

FTYPE—is a character variable that indicates whether the decision variable is a withdrawal or injection site. If FTYPE is W, the site is used for withdrawal; if FYPTE is I, the site is used for injection. If either withdrawal or injection is allowed at the site, two decision variables must be defined for the site, one for withdrawal (that is, with FTYPE=W) and one for injection (FTYPE=I).

FSTAT—is a character variable that indicates whether the decision variable will be considered in the management problem. If FSTAT is Y, the decision variable is available; if FSTAT is N, the decision variable is unavailable. If the decision variable is unavailable, then no withdrawal or injection will be calculated at the decision-variable location. For linear-optimization problems, FSTAT can be used to remove a well from the candidate set of decision variables without having to recalculate the response matrix (in this case, IRM=0; see instructions in the section "Solution and Output-Control Parameters (SOLN) File" below).

WSP—is a character string up to 120 characters long that indicates the stress periods associated with decision variable FVNAME. A single flow rate will be determined by GWM for all the stress periods included in WSP. The string must not contain any blank spaces. Multiple stress periods are indicated by colons (:) or hyphens (-). For example,

    1 indicates that stress period 1 is the only stress period associated with the decision variable,

    1:3 indicates that the flow rate is the same for stress periods 1 and 3 (but not 2), and

    1-12 indicates that the flow rate is the same for stress periods 1 through 12.

RATIO—is a real variable. RATIO is the fraction of the total flow rate for decision variable FVNAME that is distributed to cell LAY, ROW, COL. The sum of the RATIO values must equal 1.0 for all of the NC cells specified for FVNAME; if the sum does not equal 1.0, GWM calculates the fraction for each cell by dividing the RATIO value specified for each cell by the sum of the RATIO values specified for all cells within FVNAME.

EVNAME—is a character variable up to 10 characters long that is a unique name designated for the external decision variable. Each name must be unique (that is, the same name cannot be used for more than one variable, or in more than one model). No spaces are allowed in the name. The end of the name is designated by a blank space.

ETYPE—is a character variable that indicates whether the external variable is a source (import) or sink (export) of water. If ETYPE is IM, the variable is a source (import) of water; if ETYPE is EX, the variable is a sink (export) of water. Both types of external variables can be used in a management problem.

ESP—is a character string up to 120 characters long that indicates the stress periods associated with external variable EVNAME. A single flow rate will be determined by GWM for all the stress periods included in ESP. The string must not contain any blank spaces. Multiple stress periods are indicated by colons (:) or hyphens (-). For example,

    1 indicates that stress period 1 is the only stress period associated with the decision variable,

    1:3 indicates that the flow rate for the external variable is the same for stress periods 1 and 3 (but not 2), and

    1-12 indicates that the flow rate for the external variable is the same for stress periods 1 through 12.

BVNAME—is a character variable up to 10 characters long that is a unique name designated for the binary decision variable. Each name must be unique over the parent model and all child models. The use of BVNAME, NDV, and BVLIST allows the user to associate one or more FVNAME or EVNAME decision variables with a single binary-variable identifier. For example, the user may want to define 12 decision variables as the monthly withdrawal rates at a single well site. If any one of the 12 decision variables is selected in the optimal solution, then an installation cost associated with the binary variable for the well site must be incurred.

NDV—is an integer variable equal to the number of flow-rate or external decision variables associated with BVNAME.

BVLIST—is a list of the flow-rate and external decision variables associated with binary variable BVNAME. The list is drawn from the character names of these variables, FVNAME and EVNAME, defined in records 3a and 4 in any parent or child **DECVAR** file. Each character variable in the list must be separated by a space, and there must be a total of NDV variables listed. The list can extend over multiple lines; a space followed by the character "&" at the end of an input line instructs GWM to read the following line as a continuation of the list. The list can include any combination of decision variables, irrespective of well-site locations, stress period, or grid on which the decision variable has been defined.

Objective Function (**OBJFNC**) File

This file is used to define the objective function that is to be maximized or minimized and the coefficients for each decision variable in the objective function. When the LGR capability is used, the **OBJFNC** file must be defined in the parent-model **GWM** file. The GWM run will terminate if an **OBJFNC** file is referenced in a child-model **GWM** file. The OBJFNC file may reference decision variables defined for any model. Note that it is not necessary to include all decision variables in the objective function.

The **OBJFNC** file includes six input items:

0. [#Text]

Item 0 is optional—# must appear in column 1. Item 0 can be repeated multiple times.

1. IPRN

2. OBJTYP   FNTYP

3. NFVOBJ   NEVOBJ   NBVOBJ

4. The following record is repeated for each of the NFVOBJ flow-rate decision variables:
    FVNAME   FVOBJC

5. The following record is repeated for each of the NEVOBJ external decision variables:
    EVNAME   EVOBJC

6. The following record is repeated for each of the NBVOBJ binary decision variables:
    BVNAME   BVOBJC

The variables are defined as follows:

Text—is a character variable up to 199 characters long that starts in column 2. Any characters can be included in Text. Lines beginning with # are restricted to these first lines of the file. Text is printed when the file is read.

IPRN— is an integer variable that describes the amount of output that is written to the **GWM OUT** file. IPRN must be specified as either 0 or 1. When IPRN equals 0, a minimum amount of information about the objective function is written to the GWM output file; when IPRN equals 1, detailed information about the objective function is written to the GWM output file.

OBJTYP—is a character variable used to define whether the objective is to maximize or minimize the objective function. OBJTYP must be defined as either MIN (for minimize) or MAX (for maximize).

FNTYP—is a character variable used to define the type of objective function. Currently, only one type of function is available, which is WSDV for weighted sum of decision variables. The user must specify WSDV on the input record.

NFVOBJ—is an integer variable equal to the number of flow-rate decision variables in the objective function and must have a value less than or equal to NFVAR specified in the decision-variables file.

NEVOBJ—is an integer variable equal to the number of external decision variables in the objective function and must have a value less than or equal to NEVAR specified in the decision-variables file.

NBVOBJ—is an integer variable equal to the number of binary decision variables in the objective function and must have a value less than or equal to NBVAR specified in the decision-variables file.

FVNAME—is a character variable up to 10 characters long that is one of the flow-rate decision-variable names defined for any parent or child model. Each of the FVNAME variables listed must be defined in a parent or child **DECVAR** file. A flow-rate decision-variable name can be listed only once in the **OBJFNC** file.

FVOBJC—is a real variable that is a coefficient associated with each flow-rate decision variable FVNAME. For example, FVOBJC could represent the cost per unit volume of water withdrawn or injected at the management site.

EVNAME— is a character variable up to 10 characters long that is one of the external decision-variable names defined for any parent or child model. Each of the EVNAME variables listed must be defined in a parent or child **DECVAR** file. An external decision-variable name can be listed only once in the **OBJFNC** file.

EVOBJC—is a real variable that is a coefficient associated with each external decision variable EVNAME. For example, EVOBJC could represent the cost per unit volume of water associated with the external variable.

BVNAME— is a character variable up to 10 characters long that is one of the binary decision-variable names defined for any parent or child model. Each of the BVNAME variables listed must be defined in a parent or child **DECVAR** file. A binary-variable name can be listed only once in the **OBJFNC** file.

BVOBJC—is a real variable that is a coefficient associated with each binary variable BVNAME. For example, BVOBJC could represent the cost for installation of the management site. Most often, the coefficients will be positive when OBJTYP is MIN and negative when OBJTYP is MAX. This will ensure that the binary variables are active only when their associated flow-rate and external decision variables are active.

Constraint Files

Four general types of constraints can be specified in a GWM problem: constraints on the lower and upper bounds on the decision variables themselves by use of the **VARCON** file, linear-summation constraints by use of the **SUMCON** file, hydraulic head constraints by use of the **HEDCON** file, and streamflow constraints by use of the **STRMCON** file. For GWM runs that do not use the LGR capability, a **VARCON** file must be specified for each management problem; for GWM runs that use the LGR capability, a **VARCON** file must be specified for each model that contains decision variables. The linear-summation, head, and streamflow constraint files are optional. Each of these four types of constraint files is described below.

Decision-Variable Constraints (**VARCON**) File

The decision-variable constraints file is used to define lower and upper bounds for the flow-rate and external decision variables and the reference flow rates to be used in the first groundwater-flow run by GWM. Records must be specified for all NFVAR and NEVAR decision variables defined in the corresponding **DECVAR** file for a model. Reference flow rates (input variable FVREF) must be specified for each flow-rate decision variable when either drawdown constraints or streamflow-depletion constraints are used in a GWM run (see discussions on pages 15 and 31 in Ahlfeld and others, 2005); reference flow rates also may be used to calculate base conditions for the calculation of the response matrix (see discussion of variable IBASE in the **SOLN** file, and on pages 31–32 in Ahlfeld and others, 2005).

The **VARCON** file includes three input items:

0.   [#Text]

Item 0 is optional—# must appear in column 1. Item 0 can be repeated multiple times.

1. IPRN

2. The following record is read for each of the NFVAR decision variables:
   FVNAME   FVMIN   FVMAX   FVREF

3. The following record is read for each of the NEVAR decision variables:
   EVNAME   EVMIN   EVMAX

   The variables are defined as follows:

Text—is a character variable up to 199 characters long that starts in column 2. Any characters can be included in Text. Lines beginning with # are restricted to these first lines of the file. Text is printed when the file is read.

IPRN— is an integer variable that describes the amount of output that is written to the GWM **OUT** file. IPRN must be specified as either 0 or 1. When IPRN equals 0, a minimum amount of information about the decision-variable constraints is written to the GWM output file; when IPRN equals 1, detailed information about the decision-variable constraints is written to the GWM output file.

FVNAME—is a character variable up to 10 characters long that is one of the flow-rate decision-variable names defined for any parent or child model. Each of the FVNAME variables listed must be defined in a parent or child **DECVAR** file. A flow-rate decision-variable name can only be listed once in the **VARCON** file.

FVMIN and FVMAX—are real variables that are equal to the minimum (FVMIN) and maximum (FVMAX) flow rates allowed for the decision variable. Values greater than or equal to 0 must be specified for FVMIN and FVMAX; the specification of FTYPE given in the **DECVAR** file indicates whether the pumping rates are withdrawal or injection rates. FVMIN must be less than or equal to FVMAX. Note that a nonzero value of FVMIN implies that the decision variable has been associated with a binary variable in the **DECVAR** file. If the decision variable is not associated with a binary variable, then the nonzero value of FVMIN is ignored by GWM, and FVMIN is set to zero. The user can specify a nonzero lower bound for a flow-rate decision variable not associated with a binary variable by use of a linear-summation constraint (see description of **SUMCON** file).

FVREF—is a real variable equal to the flow rate for the decision variable that is used by GWM to calculate the reference values of the state variables. These include heads at drawdown-constraint locations if drawdown constraints are used and reference streamflows at streamflow-constraint locations. FVREF also may be used to calculate base conditions for the calculation of the response matrix. If no value is entered for FVREF, it is assigned a value of 0.

EVNAME— is a character variable up to 10 characters long that is one of the external decision-variable names defined for any parent or child model. Each of the EVNAME variables listed must be defined in a parent or child **DECVAR** file. An external decision-variable name can only be listed once in the **VARCON** file.

EVMIN and EVMAX—are real variables that are equal to the minimum (EVMIN) and maximum (EVMAX) flow rates allowed for the external decision variable. Values greater than or equal to 0 must be specified for EVMIN and EVMAX; the specification of ETYPE given in the **DECVAR** file indicates whether the flow rates for the external variable are imported or exported flow rates. EVMIN must be less than or equal to EVMAX. Note that a nonzero value of EVMIN implies that the decision variable has been associated with a binary variable in the **DECVAR** file. If the decision variable is not associated with a binary variable, then the nonzero value of EVMIN is ignored by GWM, and EVMIN is set to zero. The user can specify a nonzero lower bound for an external decision variable not associated with a binary variable by use of a summation constraint (see description of **SUMCON** file).

Linear-Summation Constraints (**SUMCON**) File

The linear-summation constraints file is used to define linear relations among decision variables. If the LGR capability is active and linear-summation constraints are used, then the **SUMCON** file must be listed in the parent-model GWM file. The GWM run will terminate if a **SUMCON** file is referenced in a child-model GWM file. The **SUMCON** file may include constraints that reference decision variables defined for any model.

The **SUMCON** file includes three input items:

0. [#Text]
Item 0 is optional—# must appear in column 1. Item 0 can be repeated multiple times.

1. IPRN

2. SMCNUM

3a. Records 3a and 3b are read for each of the SMCNUM constraints:
SMCNAME   NTERMS   TYPE   RHS

3b. The following record is repeated once for each of NTERMS terms specified in record 3a:
GVNAME   GVCOEFF

The variables are defined as follows:

Text—is a character variable up to 199 characters long that starts in column 2. Any characters can be included in Text. Lines beginning with # are restricted to these first lines of the file. Text is printed when the file is read.

IPRN— is an integer variable that describes the amount of output that is written to the GWM **OUT** file. IPRN must be specified as either 0 or 1. When IPRN equals 0, a minimum amount of information about the summation constraints is written to the GWM output file; when IPRN equals 1, detailed information about the summation constraints is written to the GWM output file.

SMCNUM—is an integer variable equal to the number of summation constraints defined in the file.

SMCNAME—is a character variable up to 10 characters long that is a unique name designated for the constraint. No spaces are allowed in the name. The end of the name is designated by a blank space.

NTERMS—is an integer variable equal to the number of terms on the left-hand side of the constraint. All of the terms are combined to form the left-hand side of the constraint.

TYPE—is a character variable used to specify the type of constraint. Three options are allowed:
LE indicates that the left-hand side of the constraint is less than or equal to the right-hand side of the constraint,
GE indicates that the left-hand side of the constraint is greater than or equal to the right-hand side of the constraint, and
EQ indicates that the left-hand and right-hand sides of the constraint are equal.

RHS—is a real variable equal to the value of the right-hand side of the constraint.

GVNAME—is a character variable up to 10 characters long that is one of the decision-variable names defined by a FVNAME, EVNAME, or BVNAME variable in any of the **DECVAR** files specified for the management problem. Any combination of flow-rate, external, and binary decision variables may be in a constraint. The user must ensure that the variables included are logically consistent.

GVCOEFF—is a real variable equal to the value of the coefficient associated with variable GVNAME. The user must ensure that a consistent set of units is used for all GVCOEFF and RHS terms.

Head Constraints (**HEDCON**) File

The head-constraints file is used to define head constraints at model cells. These include upper and lower bounds on heads, drawdowns, head differences between two cells, and gradients between two cells. If the LGR capability is being used, head constraints must be on the grid with which the **HEDCON** file is associated; head difference and gradient constraints between two cells may not cross the interface between grids.

The **HEDCON** file includes six input items:

0. [#Text]
Item 0 is optional—# must appear in column 1. Item 0 can be repeated multiple times.

1. IPRN

2. NHB   NDD   NDF   NGD

3. The following record is read for each of the NHB head-bound constraints:
HBNAME   LAYH   ROWH   COLH   TYPH   BND   NSP

4. The following record is read for each of the NDD drawdown constraints:
DDNAME   LAYD   ROWD   COLD   TYPD   BND   NSP

5. The following record is read for each of the NDF head-difference constraints:
HDIFNAME   LAY1   ROW1   COL1   LAY2   ROW2   COL2   HD   NSP

6. The following record is read for each of the NGD gradient constraints:
GRADNAME   LAY1   ROW1   COL1   LAY2   ROW2   COL2   LEN   GRAD   NSP

The variables are defined as follows:

Text—is a character variable up to 199 characters long that starts in column 2. Any characters can be included in Text. Lines beginning with # are restricted to these first lines of the file. Text is printed when the file is read.

IPRN— is an integer variable that describes the amount of output that is written to the GWM **OUT** file. IPRN must be specified as either 0 or 1. When IPRN equals 0, a minimum amount of information about the head constraints is written to the GWM output file; when IPRN equals 1, detailed information about the head constraints is written to the GWM output file.

NHB—is an integer variable equal to the number of head-bound constraints that need to be satisfied in the management model.

NDD—is an integer variable equal to the number of drawdown constraints that need to be satisfied in the management model.

NDF—is an integer variable equal to the number of head-difference constraints that need to be satisfied in the management model.

NGD—is an integer variable equal to the number of gradient constraints that need to be satisfied in the management model.

Head-bound constraints:

HBNAME—is a character variable up to 10 characters long that is a unique name designated for the head-bound constraint. No spaces are allowed in the name. The end of the name is designated by a blank space.

LAYH, ROWH, and COLH—are integer variables equal to the layer, row, and column number of the model cell in which the head-bound constraint is located.

TYPH—is a character variable used to specify the type of head bound. Two options are allowed:
LE indicates that head calculated by the model must be less than or equal to the value specified by BND, and

GE indicates that head calculated by the model must be greater than or equal to the value specified by BND.

BND—is a real variable equal to the specified upper or lower bound on head at the model cell at the end of the stress period (see fig. 2A in Ahlfeld and others, 2005).

NSP—is an integer variable that indicates the stress period during which the constraint is imposed. If the constraint is imposed over multiple stress periods, then a separate record must be provided for each stress period.

Drawdown constraints:

DDNAME—is a character variable up to 10 characters long that is a unique name designated for the drawdown constraint. No spaces are allowed in the name. The end of the name is designated by a blank space.

LAYD, ROWD, COLD—are integer variables equal to the layer, row, and column number of the model cell in which the drawdown constraint is located.

TYPD—is a character variable used to specify the type of drawdown bound. Two options are allowed:
LE indicates that drawdown calculated by the model must be less than or equal to the value specified by BND, and

GE indicates that drawdown calculated by the model must be greater than or equal to the value specified by BND.

BND—is a real variable equal to the specified upper or lower bound on drawdown at the model cell at the end of the stress period (see fig. 2B in Ahlfeld and others, 2005).

NSP—was defined for record 3.

Head-difference constraints:

HDIFNAME—is a character variable up to 10 characters long that is a unique name designated for the head-difference constraint. No spaces are allowed in the name. The end of the name is designated by a blank space.

LAY1, ROW1, and COL1—are integer variables equal to the layer, row, and column number of the model cell corresponding to the first location, $(i, j, k)_1$, in the head-difference constraint (see fig. 2C in Ahlfeld and others, 2005).

LAY2, ROW2, and COL2—are integer variables equal to the layer, row, and column number of the model cell corresponding to the second location, $(i, j, k)_2$, in the head-difference constraint (see fig. 2C in Ahlfeld and others, 2005). GWM requires that the head at the second location be lower than the head at the first location by an amount of at least HD. The constraint will impose the requirement that head at the second location be lower than the head at the first location by an amount of at least HD.

HD—is a real variable equal to the specified difference in heads between the first and second model cells at the end of the stress period (see fig. 2C in Ahlfeld and others, 2005).

NSP—was defined for record 3.

Gradient constraints:

GRADNAME—is a character variable up to 10 characters long that is a unique name designated for the gradient constraint. No spaces are allowed in the name. The end of the name is designated by a blank space.

LAY1, ROW1, and COL1—are integer variables equal to the layer, row, and column number of the model cell corresponding to the first location, $(i, j, k)_1$, in the gradient constraint (see fig. 2D in Ahlfeld and others, 2005).

LAY2, ROW2, and COL2—are integer variables equal to the layer, row, and column number of the model cell corresponding to the second location, $(i, j, k)_2$, in the gradient constraint (see fig. 2D in Ahlfeld and others, 2005). GWM requires that the head at the second location be lower than the head at the first location. The constraint will impose the requirement that head at the second location be lower than the head at the first location.

LEN—is a real variable equal to the distance between the first and second model cells (Δx shown in fig. 2D in Ahlfeld and others, 2005).

GRAD—is a real variable equal to the minimum gradient between the first and the second model cells at the end of the stress period (see fig. 2D in Ahlfeld and others, 2005).

NSP—was defined for record 3.

Streamflow Constraints (**STRMCON**) File

The streamflow constraints (**STRMCON**) file is used to define streamflow and streamflow-depletion constraints. Either the STR (Prudic, 1989) or SFR (Prudic and others, 2004; Niswonger and Prudic, 2005) Streamflow-Routing Packages may be used to simulate streamflow, but both packages cannot be used on a grid simultaneously. The file types specified in the MODFLOW **NAME** file indicate which package is being used to simulate streamflow; if GWM detects that both STR and SFR have been made active on a grid in the **NAME** file, the code will stop, and an error message will be printed to the GWM **OUT** file.

The input structure of the **STRMCON** file is the same regardless of which streamflow-routing package is being used. When the LGR capability is used, the constraint locations must reference locations on the grid associated with the **STRMCON** file because streams cannot cross a grid interface.

The **STRMCON** file includes four input items:

0. [#Text]
Item 0 is optional—# must appear in column 1. Item 0 can be repeated multiple times.

1. IPRN

2. NSF   NSD

3. The following record is read for each of the NSF streamflow constraints:
    SFNAME   SEG   REACH   TYPSF   BND   NSP

4. The following record is read for each of the NSD streamflow-depletion constraints:
    SDNAME   SEG   REACH   TYPSD   BND   NSP

The variables are defined as follows:

Text—is a character variable up to 199 characters long that starts in column 2. Any characters can be included in Text. Lines beginning with # are restricted to these first lines of the file. Text is printed when the file is read.

IPRN— is an integer variable that describes the amount of output that is written to the GWM **OUT** file. IPRN must be specified as either 0 or 1. When IPRN equals 0, a minimum amount of information about the streamflow constraints is written to the GWM output file; when IPRN equals 1, detailed information about the streamflow constraints is written to the GWM output file.

NSF—is an integer variable equal to the number of streamflow constraints that need to be satisfied in the management model.

NSD—is an integer variable equal to the number of streamflow-depletion constraints that need to be satisfied in the management model.

Streamflow constraints:

SFNAME—is a character variable up to 10 characters long that is a unique name designated for the streamflow constraint. No spaces are allowed in the name. The end of the name is designated by a blank space.

SEG and REACH—are integer variables equal to the segment and reach number, as specified in the STR or SFR Packages, of the model cell in which the streamflow constraint is located.

TYPSF—is a character variable used to specify the type of streamflow constraint. Two options are allowed:
> LE indicates that streamflow at the stream site must be less than or equal to the value specified by BND, and
>
> GE indicates that streamflow at the stream site must be greater than or equal to the value specified by BND.

BND—is a real variable equal to the specified amount of streamflow allowed at the stream site at the end of the stress period.

NSP—is an integer variable that indicates the stress period during which the constraint is imposed. If the constraint is imposed over multiple stress periods, then a separate record must be provided for each stress period.

Streamflow-depletion constraints:

SDNAME—is a character variable up to 10 characters long that is a unique name designated for the streamflow-depletion constraint. No spaces are allowed in the name. The end of the name is designated by a blank space.

SEG and REACH—are integer variables equal to the segment and reach number of the model cell in which the streamflow-depletion constraint is located.

TYPSD—is a character variable used to specify the type of streamflow-depletion constraint. Two options are allowed:
> LE indicates that streamflow depletion at the stream site must be less than or equal to the value specified by BND, and
>
> GE indicates that streamflow depletion at the stream site must be greater than or equal to the value specified by BND.

BND—is a real variable equal to the specified amount of streamflow depletion allowed at the stream site at the end of the stress period.

NSP—was defined for record 3.

Solution and Output-Control Parameters (**SOLN**) File

The solution and output-control parameters file is used to define several variables that control the solution algorithm for the optimization problem and the type and amount of output that is printed to the output files. Only one **SOLN** file is defined for any GWM problem and, when the LGR capability is used, the **SOLN** file must be defined in the parent-model **GWM** file. GWM will ignore a **SOLN** file that is referenced in a child-model GWM file.

The **SOLN** file includes the following input items:

0.   [#Text]

Item 0 is optional—# must appear in column 1. Item 0 can be repeated multiple times.

1.   SOLNTYP

If SOLNTYP is NS, then these records are read:

2a.   DELTA

2b.   NSIGDIG   NPGNMX   PGFACT   CRITMFC

2c.   RMNAME   IRM

Skip to record 6a.

If SOLNTYP is MPS, then these records are read:

3a. DELTA

3b. NSIGDIG   NPGNMX   PGFACT   CRITMFC

3c. MPSNAME

Skip to record 6a.

If SOLNTYP is LP, then these records are read:

4a. IRM

4b. LPITMAX   BBITMAX

4c. DELTA

4d. NSIGDIG   NPGNMX   PGFACT   CRITMFC

4e. BBITPRT   RANGE

The following record is read if IRM equals 0, 1, or 3:

4f. RMNAME1

The following record is read if IRM equals 4 or 5:

4g. RMNAME1   RMNAME2

Skip to record 6a.

If SOLNTYP is SLP, then these records are read:

5a. SLPITMAX   LPITMAX   BBITMAX

5b. SLPVCRIT   SLPZCRIT   DINIT   DMIN   DSC

5c. NSIGDIG   NPGNMX   PGFACT   AFACT   NINFMX   CRITMFC

5d. SLPITPRT   BBITPRT   RANGE

6a. IBASE

If IBASE equals 1, the following record is read for each of the NFVAR decision variables:

6b. FVNAME   FVBASE

The variables are defined as follows:

Text—is a character variable up to 199 characters long that starts in column 2. Any characters can be included in Text. Lines beginning with # are restricted to these first lines of the file. Text is printed when the file is read.

SOLNTYP—is a character variable specified as either NS, MPS, LP, or SLP:

NS indicates that no solution to the management formulation will be found by GWM. GWM will calculate the response matrix and write it to the response-matrix file specified by RMNAME in record 2c; GWM will then stop.

MPS indicates that no solution to the management formulation will be found by GWM. GWM will write the management formulation in MPS (Mathematical Programming System) format to the file specified by MPSNAME in record 3c; GWM will then stop.

LP (with or without binary variables) indicates that the optimization formulation is to be solved by using linear programming, or, for problems with binary variables, linear programming and the branch and bound method. A SOLNTYP equal to LP is normally used if the flow model contains linear features only (see the section "Solution of Ground-Water Management Problems with GWM" in Ahlfeld and others, 2005, for a discussion of linear and nonlinear features). Alternatively, the use of LP allows the user to force GWM to apply a linear solution even if nonlinear features are present. In that case, the management formulation will be solved with a single response matrix. This option should be used carefully and may lead to inaccurate results if the management problem has a significant nonlinear response.

SLP (with or without binary variables) indicates that the optimization formulation is to be solved by sequential (iterative) linear programming or, for problems with binary variables, sequential linear programming and the branch and bound method.

SOLNTYP is NS:

DELTA—is a real variable equal to the perturbation parameter δ^0 (see eq. 63 in Ahlfeld and others, 2005) used to determine the response matrix. DELTA is multiplied by FVMAX for each flow-rate decision variable to determine each perturbation value. A positive value of DELTA implies a forward-difference calculation of the response coefficient (that is, an increase in flow rate), whereas a negative value implies a backward-difference calculation (that is, a decrease in flow rate). If the flow model has a linear response to pumping, then the value assigned to DELTA has negligible significance, and values between 0.1 and 1.0 are commonly used. If the flow model has nonlinear responses, then smaller values of DELTA are preferred, commonly 0.001 to 0.1. See "Solution of Ground-Water Management Problems with GWM" in Ahlfeld and others (2005) for further discussion of DELTA.

NSIGDIG—is an integer variable equal to a lower limit on the number of significant digits in response-matrix entries. For each entry in a column of the response matrix, the ratio of the difference in observed state to the HCLOSE variable is computed. If the largest ratio in the column has fewer than NSIGDIG significant digits, the perturbation is considered to have failed. Values of NSIGDIG of 1 to 3 are often adequate for small problems. Higher values should be used as problems become more complex. See "Description of Selected Conventions, Options, and Variables in GWM" in Ahlfeld and others (2005, p. 32–33) for further discussion of NSIGDIG.

NPGNMX—is an integer variable that controls the automatic reexecution of the flow process when it fails. When failure is detected, automatic resetting of flow-rate decision-variable values may produce a successful solution. Failure may occur during either base or perturbation flow-process runs. For perturbation runs, the resetting is controlled by PGFACT, whereas for base runs, the resetting is controlled by AFACT (see record 5c). The value of NPGNMX is the maximum number of attempts to achieve a successful flow-process run. When NPGNMX is set to 0, testing of the flow-process results is eliminated except to confirm that the flow process has converged. If NPGNMX is set to 0 and GWM determines that the GWF Process has failed to converge, then the GWM run will be terminated.

PGFACT—is a real variable equal to the perturbation step-length adjustment factor used during perturbation failure. PGFACT must be greater than 0 and less than 1. The smaller the value of PGFACT, the larger the adjustment to the perturbation value. A value of 0.5 is suggested. See "Description of Selected Conventions, Options, and Variables in GWM" in Ahlfeld and others (2005, p. 33) for further discussion of PGFACT.

CRITMFC—is a real variable used to adjust the criteria for accepting convergence of a GWF Process run. For a single-grid problem, the options are:

CRITMFC = 0.0: convergence of the GWF Process will be assumed only if the applicable GWF convergence criteria (HCLOSE and/or RCLOSE) have been met at the end of each time step (this is the standard approach for the GWF Process of MODFLOW);

CRITMFC > 0.0: the results of the GWF Process will be accepted if the applicable GWF convergence criteria (HCLOSE and/or RCLOSE) have not been met, but the percent discrepancy between the inflow and outflow rates for each time step as reported in the GWF Process volumetric budget is less than or equal to the value specified for CRITMFC. For example, if CRITMFC is specified as 0.1, then GWF Process results will be accepted if the percent discrepancy between the inflow and outflow rates of the volumetric budget at the end of each time step is less than or equal to 0.1 percent;

CRITMFC < 0.0: the results of the GWF Process will be accepted regardless of the GWF convergence criteria (HCLOSE and/or RCLOSE) or the percent discrepancy.

For multiple grids, the options are:

CRITMFC = 0.0: convergence of the GWF Process will be assumed only if the HCLOSELGR and FCLOSELGR criteria have been met at the end of each time step (this is the standard approach for an LGR simulation);

CRITMFC > 0.0: the results of the GWF Process will be accepted if the HCLOSELGR or FCLOSELGR criteria have not been met, but the percent discrepancy between the inflow and outflow rates for each time step as reported in the GWF Process volumetric budget for all grids is less than or equal to the value specified for CRITMFC. For example, if CRITMFC is specified as 0.1, then GWF Process results will be accepted if the percent discrepancy between the inflow and outflow rates of the volumetric budget for all grids at the end of each time step is less than or equal to 0.1 percent;

CRITMFC < 0.0: the results of the GWF Process will be accepted regardless of the HCLOSELGR or FCLOSELGR criteria or the percent discrepancy.

RMNAME—is the filename (or pathname) to which the response matrix will be written.

IRM—is an integer variable. Its value specifies the format in which the response matrix will be written to file RMNAME (if no value is specified for IRM, a default value of 1 is assumed):

IRM = 1: the response matrix is saved to a nonformatted file;

IRM = 3: the response matrix is printed to a formatted file.

SOLNTYP is MPS:

DELTA—was defined for item 2a.

NSIGDIG, NPGNMX, PGFACT, CRITMFC—were defined for item 2b.

MPSNAME—is the filename (or pathname) to which the formulation will be written in MPS format.

SOLNTYP is LP:

IRM—is an integer variable. Its value specifies whether or not the response matrix will be calculated or read from an input file, and whether or not the response matrix will be saved:

IRM = 0: An existing nonformatted file containing the response matrix will be read from the file specified by RMNAME1 in item 4f, and the LP solved.

IRM = 1: The response matrix is computed and saved to a nonformatted file specified by RMNAME1 in item 4f. The LP is then solved.

IRM = 2: The response matrix is computed and the LP solved. The response matrix is not saved or printed.

IRM = 3: The response matrix is computed and printed to a formatted file specified by RMNAME1 in item 4f. The LP is then solved.

IRM = 4: The response matrix is computed, saved to a nonformatted file specified by RMNAME1 in item 4g, and printed to a formatted file specified by RMNAME2 in item 4g. The LP is then solved.

IRM = 5: An existing nonformatted file containing the response matrix is read from the file specified by RMNAME1 in item 4g, and the LP solved. The response matrix is printed to a formatted file specified by RMNAME2 in item 4g.

LPITMAX—is an integer variable. Its value is the maximum number of iterations allowed for the linear program solver; this limit prevents the solver from iterating indefinitely if it does not converge to a solution. If the linear solver is being used, and the value of LPITMAX is reached, the program will be terminated, and the output file will indicate that the maximum number of iterations has been reached. A typical value for LPITMAX is ten times the number of constraints.

BBITMAX—is an integer variable. BBITMAX is relevant only if the management problem includes one or more binary variables; otherwise, its value is ignored. BBITMAX is the maximum number of iterations allowed for the branch and bound program solver. Each iteration consists of one solution of the linear program. If the value of BBITMAX is reached, the program will be terminated, and the output file will indicate that the maximum number of iterations has been reached.

DELTA—was defined for item 2a.

NSIGDIG, NPGNMX, PGFACT, CRITMFC—were defined for item 2b.

BBITPRT—is an integer variable that specifies whether output describing the details of the branch and bound algorithm for solving mixed binary problems will be written to the GWM **OUT** file. A value of 1 indicates that this output will be written, and a value of 0 indicates that it will not. For problems with many binary variables, values of 1 can substantially increase the size of the GWM **OUT** file. If the management problem includes no binary variables, this value will be ignored.

RANGE—is an integer variable that indicates the status of the range analysis. A value of 1 indicates that range analysis is to be done and the results written to the GWM **OUT** file. A value of 0 indicates that no range analysis is to be done. Range analysis is based on the assumption that the optimization problem is strictly linear with continuous variables. If binary variables or nonlinear responses are significant in the problem, then the range analysis may be inaccurate.

RMNAME1—is a filename (including pathname, if required) from which the response matrix will be read if IRM equals 0 or 5 and to which the response matrix will be written if IRM equals 1, 3, or 4. RMNAME1 will be a formatted file if IRM equals 3 and a nonformatted file otherwise. The type of nonformatted file is defined in the openspec.inc file distributed with GWM.

RMNAME2— is a filename (including pathname, if required) to which a response matrix will be written if IRM equals 4 or 5.

SOLNTYP is SLP:

SLPITMAX—is an integer variable. Its value is the maximum number of iterations allowed for the SLP algorithm. If the value of SLPITMAX is reached, the program will be terminated, and the output file will indicate that the maximum number of iterations has been reached.

LPITMAX—was defined for item 4b.

BBITMAX—was defined for item 4b.

SLPVCRIT—is a real variable. Its value is the convergence criterion ε_1 (see eq. 69A in Ahlfeld and others, 2005), which is the first of two termination criteria when the SLP algorithm is used. This criterion is satisfied when the absolute value of the change in the values of all flow-rate decision variables from the previous iteration to the current iteration is less than a fraction ε_1 of the magnitude of the flow-rate decision variables at the current iteration. See Ahfled and others (2005, p. 25) for guidance on specifying the value of SLPVCRIT.

SLPZCRIT—is a real variable. Its value is the convergence criterion ε_2 (see eq. 69B in Ahlfeld and others, 2005), which is the second of two termination criteria when the SLP algorithm is used. This criterion is satisfied when the absolute value of the change in the value of the objective function is less than the specified fraction ε_2 of the value of the objective function.

DINIT, DMIN, and DSC—are real variables that control the value of the perturbation variable (see eq. 72 in Ahlfeld and others, 2005) used to compute response coefficients. DINIT is the perturbation variable used for the first iteration, DMIN is the minimum perturbation variable used, and DSC is a parameter that controls the rate of change of the perturbation variable. DINIT and DMIN must have the same sign. Positive values of DINIT and DMIN imply a forward-difference calculation of the response coefficient (that is, an increase in flow rate), whereas negative values imply a backward-difference calculation (that is, a decrease in flow rate). DSC must always be positive. See Ahlfeld and others (2005, p. 27 and 32) for discussions of DINIT, DMIN, and DSC.

NSIGDIG, NPGNMX, PGFACT, CRITMFC—were defined for item 2b.

AFACT—is a real variable equal to the relaxation parameter α (see eq. 73 in Ahlfeld and others, 2005) used to determine a temporary base solution when a base run fails. AFACT controls the interpolation between the current base solution and the most recent successful base solution. AFACT must be greater than 0 and less than 1. A value close to 0 implies that the temporary base solution will be close to the current base solution, whereas a value close to 1.0 implies that the temporary base solution will be close to the previous base solution. A value of 0.5 is suggested.

NINFMX—is an integer variable that specifies the maximum number of consecutive infeasible iterations that will be accepted by the SLP algorithm before the algorithm terminates.

SLPITPRT—is an integer variable that specifies whether output describing the details of the sequential-iteration algorithm will be written to the GWM **OUT** file. The options are:

SLPITPRT = 0: No information on the progress of the SLP algorithm is written to the **OUT** file.

SLPITPRT = 1: Constraint and convergence status at each iteration of the SLP algorithm are written to the **OUT** file.

SLPITPRT = 2: The current optimal solution, constraint status, and convergence status at each iteration of the SLP algorithm are written to the **OUT** file. If GWMWFILE is active, then a new well file will be written based on the current optimal solution.

BBITPRT—was defined in item 4e.

RANGE—was defined in item 4e.

IBASE—is an integer variable equal to 0 or 1 that indicates the source for the values of the flow-rate decision variables that will be used as the base run. For problems solved using the SLP algorithm, these values are the starting point for the iterative algorithm. A value of IBASE equal to 0 indicates that the reference flow rates (FVREF) specified for each flow-rate decision variable in file VARCON will be used in the base run (and that record 6b is not necessary). A value of IBASE equal to 1 indicates that the flow rates specified for each decision variable by FVBASE in record 6b will be used to calculate the base run. See discussion about IBASE in the "Variables related to calculation of response coefficients" section of Ahlfeld and others (2005, p. 31–32).

FVNAME—is a character variable up to 10 characters long that is one of the flow-rate decision-variable names. Each name must be unique to the parent model or one of the child models (that is, the same name cannot be used in more than one model). A flow-rate decision-variable name can only be listed once in the **SOLN** file.

FVBASE—is a real variable equal to the rate for the flow-rate decision variable. These values are used by GWM to calculate the base run. If the SLP solution algorithm is used, these values are the starting point for the iterative algorithm.

Output Files

Two types of output files are always produced by a GWM-2005 run—an **OUT** file, which may be specified in the parent **GWM** file, and a **LIST** file for each parent and child model. Three additional types of files are also produced if specified by the user in the **SOLN** and **DECVAR** files—one or two files to hold response matrixes, a file to hold formulations written in MPS format, and a file to hold values of the flow-rate decision variables at the optimal solution. As described previously, the **OUT** file is used in GWM-2005 in place of the original GWM-2000 **GLOBAL** file, which was used to hold information about the GWM run. The contents of the **OUT** file are similar in nearly all aspects to those of the original **GLOBAL** file, and the user is encouraged to review the description of the contents of the file in Ahlfeld and others (2005, p. 50-52). The only substantial change to the **OUT** file occurs when the LGR capability is used, in which case the names of the input files from which the decision variables and head and streamflow constraints are read for each model are written to the **OUT** file. The **LIST** files for each parent and child model hold the output from a particular run of the GWF Process, as well as water-budget information for flows between parent and child model in the LGR solution process. Each **LIST** file is erased and generated anew each time a GWF Process run is required.

Sample Problems

Four sample problems are provided to demonstrate some of the capabilities of GWM-2005, to provide examples of the selection of input parameters for solution of management problems with LGR, and to demonstrate that the link between the GWM and LGR codes is working as intended. The first sample problem describes the use of the new SFR streamflow-routing option for GWM-2005 (with LGR inactive) and the remaining three problems demonstrate use of the LGR capability. Selected input and output files are included with each of the sample problems; all input files for each of the sample problems are included in the GWM-2005 distribution package available at the USGS Internet URL provided in the Preface to this report. Before these sample problems are described, changes to the original sample problems provided in Ahlfeld and others (2005) are described.

Changes to Original Sample Problems

Minor changes were made to the sample problems described in Ahlfeld and others (2005) for the conversion of GWM-2000 to GWM-2005. First, the input files for all sample problems were modified to remove the **GLOBAL** file specification from the **NAME** file and to add the name of a GWM **OUT** file to the **GWM** file. Second, the SEAWATER problem distributed with GWM-2000 included a Sensitivity (SEN) Process input file to set the value of recharge; use of the **SEN** file overrode the value specified for recharge in the RCH Package input file. The **SEN** input file was added to demonstrate the ability of GWM to read MODFLOW parameters. However, the SEN Process is not available in MODFLOW-2005. Instead, users can specify MODFLOW parameters by use of the Parameter Value (**PVAL**) file. As a result, the SEAWATER problem now includes a **PVAL** file. The output file from the sample problems distributed with GWM-2005 can still be compared to those in Ahlfeld and others (2005), except that the information that was previously written to the **GLOBAL** file from the GWF Process is now written to the GWM **OUT** file.

An additional change also was made to the SUPPLY sample problem. The SUPPLY problem in the original GWM documentation (Ahlfeld and others, 2005, p. 82) describes a transient water-supply problem in which total groundwater withdrawals over a 3-year period are limited by the amount of streamflow depletion allowed in two streams in hydraulic connection with the simulated aquifer. Experience with the sample problem indicated that there could be numerical-stability issues associated with its solution that resulted from streamflow-depletion constraints that were nearly redundant from one year to the next. Because of these numerical-stability issues, the sample problem was revised, and is now referred to as the SUPPLY2 problem. The changes to the problem were as follows: (1) flow-rate decision variable $Q3$ was dropped from the formulation, (2) water-supply demands in the second year were removed, and (3) streamflow-depletion constraints in the first two years were removed. These changes resulted in a reduction in the numbers of (1) candidate flow-rate decision variables from 8 to 7, (2) specified water-supply demand constraints (which are specified by using the summation constraints option of GWM) from 24 to 16, and (3) streamflow-depletion constraints from 36 to 12. The revised solution obtained using GWM is nearly identical to the solution to the original SUPPLY problem, although decision variable $Q3$ is no longer active in the solution. The value of the objective function at the optimal solution is $53,022 for both the original and revised problems.

SUPPLY2 Sample Problem with Streamflow-Routing (SFR) Package

The SUPPLY2 sample problem was further revised to demonstrate use of the SFR Package in place of the STR Package to simulate groundwater/surface-water interactions and streamflow routing. The only changes that were required to the input files for the sample problem were that (1) the STR file type was replaced by the SFR file type in the MODFLOW **NAME** file and (2) an STR input file was replaced by an SFR input file. The SFR input file prepared for the sample problem is shown in figure 6.

Although there are several differences in the structure of the input files between the STR and SFR Packages, the type of information that is specified for both packages is generally the same. The two differences that affect the sample problem relate to the specification of stream-channel slope and streambed conductance in the two input files. In SFR, stream-channel slope, which affects the calculation of stream depth in each simulated stream reach, is not specified, as it is for the STR Package. Instead, stream-channel slope is calculated by SFR on the basis of streambed elevations and stream-channel lengths specified for each stream segment and reach, respectively. Also, streambed conductance is not specified directly in SFR, as it is for STR. Instead, the user must specify the hydraulic conductivity (K) and thickness (M) of the streambed in each stream segment, the width (W) of the stream channel for each stream segment, and the length (L) of the stream channel in each stream reach. SFR then calculates the streambed conductance for each stream reach on the basis of the relation KLW/M.

Because it was not possible to specify streambed elevations and stream-channel lengths in the SFR input file in such a way that calculated stream-channel slopes were exactly equal to those specified in the STR input file, there were small differences in the computed values of streamflow, stream depth, and groundwater discharge to the simulated streams between the STR and SFR simulations. Nevertheless, the differences between the optimal solutions to the management formulation calculated by GWM by use of the STR and SFR Packages were small. The value of the objective function at the optimal solution is $53,022 with the STR Package and $53,028 with the SFR Package. All seven flow-rate decision variables and all four external decision variables are active in each of the STR and SFR solutions, and the optimal values calculated for each of the decision variables in the two solutions are virtually identical.

DEWATER-LGR

This sample problem, which represents a steady-state dewatering problem for a construction site, is derived from the DEWATER sample problem described in the original GWM manual (Ahlfeld and others, 2005). The objective of the groundwater-management problem is to minimize the cost of withdrawing groundwater to lower heads to an elevation of 50 ft so that footings can be installed in the area shown in figure 7. Wells at the construction site will be pumped for 1,000 days. The aquifer at the site is confined and is simulated by a single model layer that is 3,000 ft long and 2,000 ft wide. The original sample problem has been modified by refining the grid in the area of the construction site with three alternative levels of horizontal refinement—1:1, 3:1, and 7:1. The alternative grids are used to do a sensitivity analysis of grid refinement and to compare results with those for the original, unrefined grid.

The parent model consists of 20 rows and 30 columns (fig. 7), and each grid cell is 100 ft by 100 ft. The area of refinement (that is, the extent of the child model) extends from row 4, column 10 to row 16, column 19 of the parent model (fig. 7). The child model with refinement of 1:1 has 13 rows and 10 columns, and each cell

Figure 6. Input file for Streamflow-Routing (SFR) Package for SUPPLY2 sample problem.

```
#SUPPLY2 Sample Problem using SFR                          Record 0
40    3  0  0  128390.  0.0001   -1  81                    Record 1
    1  13   1   1    1                200.0                Record 2
    1  13   2   1    2                200.0
    1  14   3   1    3                200.0
    1  15   4   1    4                200.0
    1  16   5   1    5                200.0
    1  16   6   1    6                200.0
    1  16   7   1    7                200.0
    1  16   8   1    8                200.0
    1  16   9   1    9                200.0
    1  16  10   1   10                200.0
    1  16  11   1   11                200.0
    1  16  12   1   12                200.0
    1  15  13   1   13                200.0
    1  14  14   1   14                200.0
    1  13  15   1   15                200.0
    1  12  16   1   16                200.0
    1  12  17   1   17                200.0
    1  12  18   1   18                200.0
    1  12  19   1   19                200.0
    1  13  20   1   20                200.0
    1  14  21   1   21                200.0
    1  14  22   1   22                200.0
    1   9  30   2    1                200.0
    1   9  29   2    2                200.0
    1   9  28   2    3                200.0
    1   9  27   2    4                200.0
    1  10  26   2    5                200.0
    1  11  25   2    6                200.0
    1  12  24   2    7                200.0
    1  13  23   2    8                200.0
    1  14  22   2    9                200.0
    1  14  22   3    1                200.0
    1  15  23   3    2                200.0
    1  15  24   3    3                200.0
    1  15  25   3    4                200.0
    1  15  26   3    5                200.0
    1  16  27   3    6                200.0
    1  17  28   3    7                200.0
    1  18  29   3    8                200.0
    1  18  30   3    9                200.0
    3   0   0                                              Record 5: St.Prd. 1
    1   1   3   0 100000.0  0.0   0.0  0.0  0.05           Record 6a: Seg. 1
        5.0        1.0      48.50      20.0       3.0      Record 6b
        5.0        1.0      38.00      20.0       3.0      Record 6c
    2   1   3   0  50000.0  0.0   0.0  0.0  0.05           Record 6a: Seg. 2
        5.0        1.0      40.25      20.0       3.0      Record 6b
        5.0        1.0      38.00      20.0       3.0      Record 6c
    3   1   0   0      0.0  0.0   0.0  0.0  0.05           Record 6a: Seg. 3
        5.0        1.0      38.00      20.0       3.0      Record 6b
        5.0        1.0      30.00      20.0       3.0      Record 6c
       -3        0         0                               Record 5: St.Prd. 2
    (Records for Stress Periods 3-11 deleted here.)
       -3        0         0                               Record 5: St.Prd.12
```

is 100 ft by 100 ft; the child model with refinement of 3:1 has 37 rows and 28 columns, and each cell is 33.3 ft by 33.3 ft; the child model with refinement of 7:1 has 85 rows and 64 columns, and each cell is 14.3 ft by 14.3 ft.

The parent model uses no-flow boundary conditions along the north and south boundaries of the aquifer and constant heads of 80 ft and 60 ft along the east and west boundaries of the aquifer, respectively. The transmissivity of the aquifer is 50 ft^2/d.

Each of the parent and child models consists of a **NAME** file, a **DIS** file, a **BAS** file, a **BCF** file, and a **PCG** file. Both models use the same **PCG** file, in which a head-change convergence criterion of 5.0×10^{-8} ft is specified. This small value is used to ensure that differences resulting from alternate grid refinements are not masked by roundoff error.

Linear Groundwater-Management Formulation

Seven candidate well locations, all of which are within the child model, are selected as possible locations of withdrawal (fig. 7). These seven flow-rate decision variables (named Q1, Q2, and so forth) are specified in the **DECVAR** file for the child model. Upper and lower bounds on the withdrawal rates for each well are specified in the **VARCON** file for the child model. The parent model has no **DECVAR** or **VARCON** files.

The objective function for the problem, which is specified in the **OBJFNC** file for the parent model, is to minimize the sum of withdrawals at the seven candidate well locations. Each flow-rate decision variable is assigned a coefficient (input variable FVOBJC) of 1.0 in the **OBJFNC** file. The 50-ft head criterion is imposed by defining 10 upper-bound head constraints in the **HEDCON** file for the child model at the locations shown in figure 7. The groundwater-management problem is linear because the aquifer is confined, there are no head-dependent boundary conditions, and the objective function and all of the constraints are linear. Therefore, it is

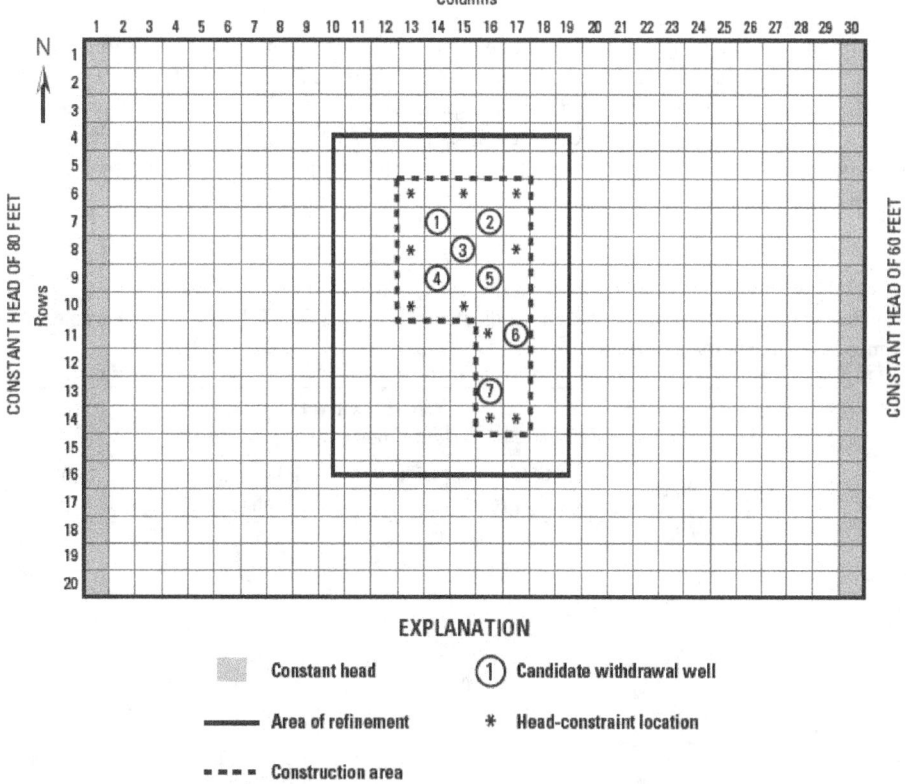

EXPLANATION

| | Constant head | ① | Candidate withdrawal well |

——— Area of refinement   *   Head-constraint location

- - - - Construction area

Figure 7. Schematic diagram showing model grid for DEWATER-LGR sample problem with local area of refinement.

solved using the LP option of GWM. The **SOLN** file is specified for the parent model. Input variables speci-fied in the **SOLN** file are identical to those used in the original sample problem described in Ahlfeld and others (2005). The LGR convergence variables specified in the LGR **CONTROL** file for all simulations are 0.4 for RELAXH and RELAXF, 5.0×10^{-5} for HCLOSELGR, and 5.0×10^{-4} for FCLOSELGR.

The value of the objective function at the optimal solution calculated by GWM-2005 for the original problem (that is, with no refinement) is 2.8657×10^6 ft^3 of water withdrawn, which is the same value obtained with GWM-2000 (Ahlfeld and others, 2005). With the LGR capability active, the value of the objective func-tion is 2.8656×10^6 ft^3 for 1:1 refinement, 2.8579×10^6 ft^3 for 3:1 refinement, and 2.8570×10^6 ft^3 for 7:1 refine-ment. In all four simulations, the same four wells were selected for pumping in the optimal solution (table 2). In addition, in all cases, four of the head constraints—those located at cells (1, 6, 13), (1, 6, 17), (1, 10, 13), and (1, 14, 17)—were binding at the optimal solution. Note in table 2 that the withdrawal rates for the 1:1 refinement simulation are the same as those calculated for the simulation with no refinement, as expected. The withdrawal rates calculated for each well for the 3:1 and 7:1 refinement ratios reflect the increased accuracy of the heads calculated at the constraint locations by use of the locally refined models; that is, the increased accuracy in heads calculated as part of the modified GWF Process with LGR results in more accurate response coefficients than those calculated without LGR, which in turn results in changes in the optimal withdrawal rates calculated by GWM. Although differences in the results among the four simulations are small for this particular problem, the results indicate that the link between the GWM and LGR codes is working properly.

Output from the GWM-2005 simulation is printed to the **OUT** file specified in the parent **GWM** file. MODFLOW **LIST** files are created for both the parent and child models. The volumetric budgets at the opti-mal solution for the parent and child models (which are provided for the 3:1 refinement simulation in the next section of the report) illustrate a number of useful budget values for the problem. First, the ultimate source of water to the wells (2,857.9668 ft^3/d as given in the volumetric budgets) is inflow from the constant heads along the parent model. This same rate of flow enters the child model from the parent model, as seen in the entry "TOTAL INTO CHILD" near the bottom of the child-model volumetric-budget information and in the entry "PARENT FLUX B.C." (boundary condition) in the parent-model volumetric-budget information. The rates of MANAGED FLOW are zero for the parent model because there are no managed wells in the parent model and 2,857.9668 ft^3/d for the child model because all of the water is withdrawn in the child model.

GWM run times for the four grid-refinement approaches increase from 0.25 seconds for the no-refine-ment simulation to 1.109 seconds for 1:1 refinement, to 2.407 seconds for the 3:1 refinement, and 11.172 sec-onds for the 7:1 refinement (on a Dell Latitude D630 laptop with an Intel Core2 Duo microprocessor running at 2.59 gigahertz and 2.00 gigabytes of random-access memory). Although these run times are short because of the relative simplicity of the problem, they indicate the proportional increase in run times that can be expected for increased grid refinement.

Table 2. Optimal withdrawal rates for the seven candidate wells in the DEWATER-LGR sample problem calculated with four grid-refinement approaches.

Well	Withdrawal rate (cubic feet per day)			
	No grid refinement	Child model with 1:1 refinement	Child model with 3:1 refinement	Child model with 7:1 refinement
Q1	1,077.39	1,077.39	1,060.10	1,058.10
Q2	78.24	78.24	96.01	98.04
Q3	0	0	0	0
Q4	768.95	768.95	768.60	768.61
Q5	0	0	0	0
Q6	0	0	0	0
Q7	941.08	941.08	933.25	932.31
Total	**2,865.66**	**2,865.66**	**2,857.97**	**2,857.06**

Selected Input and Output Files for Sample Problem

LGR CONTROL file for 3:1 refinement (*deworg3.lgr*)

```
#DEWATER-LGR sample problem using 1 child grid with refinement ratio 3:1
LGR                     ;LGR keyword
2                       ;NGRIDS
..\data\DEWATER-LGR\deworg3_par.nam        ;NAMEFILE for parent
PARENTONLY              ;GRIDSTATUS
00 00                   ;IUPBHSV,IUPBFSV
..\data\DEWATER-LGR\deworg3_chd.nam        ;NAMEFILE for child
CHILDONLY               ;GRIDSTATUS
1 -59  0   0            ;ISHFLG, IBFLG, IUCBHSV, IUCBFSV
50 0                    ;MXLGRITER, IOUTLGR
0.40000 0.4000          ;RELAXH, RELAXF    relaxation for heads and fluxes
5.0E-5 5.0E-4           ;HCLOSELGR FCLOSELGR
1 4 10                  ;NPLBEG,NPRBEG,NPCBEG
1 16 19                 ;NPLEND,NPREND,NPCEND
3                       ;NCPP   of child cells per width of parent
1                       ;NCPPL  # of child layers per parent layer
```

NAME file for parent model (*deworg3_par.nam*)

```
LIST   10    deworg3_par.lst
DIS    11    ..\data\DEWATER-LGR\deworg3_par.dis
BAS6   12    ..\data\DEWATER-LGR\deworg3_par.ba6
BCF6   13    ..\data\DEWATER-LGR\deworg3_par.bc6
PCG    14    ..\data\DEWATER-LGR\deworg3.pcg
GWM    15    ..\data\DEWATER-LGR\deworg3_par.gwm
```

GWM file for parent model (*deworg3_par.gwm*)

```
#DEWATER-LGR Sample Problem, parent GWM file
OUT      deworg3_par.out
OBJFNC   ..\data\DEWATER-LGR\deworg3_par.objfnc
SOLN     ..\data\DEWATER-LGR\deworg3_par.soln
```

Objective Function (OBJFNC) file specified for parent model (*deworg3_par.objfnc*)

```
#DEWATER-LGR Sample Problem, OBJFNC file
 1              #1-IPRN
 MIN  WSDV      #2-OBJTYP  FNTYP
 7  0  0        #3-NFVOBJ  NEVOBJ   NBVOBJ
 Q1  1.0        #4-FVNAME  FVOBJC
 Q2  1.0
 Q3  1.0
 Q4  1.0
 Q5  1.0
 Q6  1.0
 Q7  1.0
```

Solution and output control (SOLN) file specified for parent model (*deworg3_par.soln*)

```
#DEWATER-LGR Sample Problem, SOLN file
#June, 2009
 LP             #1-SOLNTYP
 2              #4a-IRM
 1000  2000     #4b-LPITMAX  BBITMAX
 0.5            #4c-DELTA
 1 10  0.5  0.0 #4d-NSIGDIG  NPGNMX  PGFACT  CRITMFC
 1  1           #4e-BBITPRT  RANGE
 0              #6a-IBASE
```

NAME file for child model (*deworg3_chd.nam*)

```
LIST  20   deworg3_chd.lst
DIS   21   ..\data\DEWATER-LGR\deworg3_chd.dis
BAS6  22   ..\data\DEWATER-LGR\deworg3_chd.ba6
BCF6  23   ..\data\DEWATER-LGR\deworg3_chd.bc6
PCG   24   ..\data\DEWATER-LGR\deworg3.pcg
GWM   25   ..\data\DEWATER-LGR\deworg3_chd.gwm
```

GWM file for child model (*deworg3_chd.gwm*)

```
#DEWATER-LGR Sample Problem, child GWM file
DECVAR  ..\data\DEWATER-LGR\deworg3_chd.decvar
VARCON  ..\data\DEWATER-LGR\deworg3_chd.varcon
HEDCON  ..\data\DEWATER-LGR\deworg3_chd.hedcon
```

Decision variable (DECVAR) file for child model (*deworg3_chd.decvar*)

```
#DEWATER-LGR Sample Problem, DECVAR file
1 0                        #1-IPRN  GWMWFILE
7 0 0                      #2-NFVAR  NEVAR  NBVAR
Q1  1   1 10  13  W  Y  1  #3a-FVNAME NC LAY ROW COL FTYPE FSTAT WSP
Q2  1   1 10  19  W  Y  1
Q3  1   1 13  16  W  Y  1
Q4  1   1 16  13  W  Y  1
Q5  1   1 16  19  W  Y  1
Q6  1   1 22  22  W  Y  1
Q7  1   1 28  19  W  Y  1
```

Decision-variable constraints (VARCON) file for child model (*deworg3_chd.varcon*)

```
#DEWATER-LGR Sample Problem, VARCON file
 1                         #1-IPRN
Q1 0.0d2  2.0d4  0.0d2     #2-FVNAME  FVMIN  FVMAX  FVREF
Q2 0.0d2  2.0d4  0.0d2
Q3 0.0d2  2.0d4  0.0d2
Q4 0.0d2  2.0d4  0.0d2
Q5 0.0d2  2.0d4  0.0d2
Q6 0.0d2  2.0d4  0.0d2
Q7 0.0d2  2.0d4  0.0d2
```

Head constraints (HEDCON) file for child model (*deworg3_chd.hedcon*)

```
#DEWATER-LGR Sample Problem, HEDCON file
 1                         #1-IPRN
10 0  0  0                 #2-NHB NDD NDF NGD
b-01  1  7 10 le 50.0 1    #3-HBNAME LAYH ROWH COLH TYPH BND NSP
b-02  1  7 16 le 50.0 1
b-03  1  7 22 le 50.0 1
b-04  1 13 10 le 50.0 1
b-05  1 13 22 le 50.0 1
b-06  1 19 10 le 50.0 1
b-07  1 19 16 le 50.0 1
b-08  1 22 19 le 50.0 1
b-09  1 31 19 le 50.0 1
b-10  1 31 22 le 50.0 1
```

Volumetric budget for parent model at optimal solution (from parent-model LIST file, *deworg3_par.lst*)

```
VOLUMETRIC BUDGET FOR ENTIRE MODEL AT END OF TIME STEP  1 IN STRESS PERIOD    1
--------------------------------------------------------------------------------


     CUMULATIVE VOLUMES      L**3        RATES FOR THIS TIME STEP      L**3/T
     ------------------                  ------------------------


          IN:                                 IN:
          ---                                 ---
            STORAGE =        0.0000            STORAGE =        0.0000
      CONSTANT HEAD =  2857966.8000      CONSTANT HEAD =     2857.9668
       MANAGED FLOW =        0.0000       MANAGED FLOW =        0.0000
    PARENT FLUX B.C. =       0.0000    PARENT FLUX B.C. =       0.0000

           TOTAL IN =  2857966.8000           TOTAL IN =     2857.9668

          OUT:                                OUT:
          ----                                ----
            STORAGE =        0.0000            STORAGE =        0.0000
      CONSTANT HEAD =        0.0000      CONSTANT HEAD =        0.0000
       MANAGED FLOW =        0.0000       MANAGED FLOW =        0.0000
    PARENT FLUX B.C. =  2857966.7986    PARENT FLUX B.C. =    2857.9668

          TOTAL OUT =  2857966.7986          TOTAL OUT =     2857.9668

          IN - OUT =     1.3314E-03          IN - OUT =     1.3314E-06

  PERCENT DISCREPANCY =        0.00    PERCENT DISCREPANCY =        0.00



          TIME SUMMARY AT END OF TIME STEP   1 IN STRESS PERIOD    1
                    SECONDS      MINUTES      HOURS       DAYS        YEARS
                 -----------------------------------------------------------
    TIME STEP LENGTH 8.64000E+07 1.44000E+06  24000.      1000.0      2.7379
  STRESS PERIOD TIME 8.64000E+07 1.44000E+06  24000.      1000.0      2.7379
         TOTAL TIME 8.64000E+07 1.44000E+06  24000.      1000.0      2.7379
```

Volumetric budget for child model at optimal solution (from child-model LIST file, *deworg3_chd.lst*)

```
VOLUMETRIC BUDGET FOR ENTIRE MODEL AT END OF TIME STEP  1 IN STRESS PERIOD    1
--------------------------------------------------------------------------------


     CUMULATIVE VOLUMES      L**3        RATES FOR THIS TIME STEP      L**3/T
     ------------------                  ------------------------


          IN:                                 IN:
          ---                                 ---
            STORAGE =        0.0000            STORAGE =        0.0000
      CONSTANT HEAD =  2857966.7973      CONSTANT HEAD =     2857.9668
       MANAGED FLOW =        0.0000       MANAGED FLOW =        0.0000

           TOTAL IN =  2857966.7973           TOTAL IN =     2857.9668
```

```
          OUT:                                        OUT:
          ----                                        ----
              STORAGE =        0.0000                     STORAGE =        0.0000
        CONSTANT HEAD =        0.0000               CONSTANT HEAD =        0.0000
         MANAGED FLOW =  2857966.8045                MANAGED FLOW =     2857.9668


            TOTAL OUT =  2857966.8045                   TOTAL OUT =     2857.9668


              IN - OUT =   -7.2494E-03                    IN - OUT =   -7.2494E-06


    PERCENT DISCREPANCY =          0.00       PERCENT DISCREPANCY =          0.00



            TIME SUMMARY AT END OF TIME STEP   1 IN STRESS PERIOD    1
                         SECONDS      MINUTES      HOURS       DAYS        YEARS
                         ----------------------------------------------------------------
        TIME STEP LENGTH 8.64000E+07 1.44000E+06  24000.     1000.0      2.7379
     STRESS PERIOD TIME 8.64000E+07 1.44000E+06  24000.     1000.0      2.7379
            TOTAL TIME 8.64000E+07 1.44000E+06  24000.     1000.0      2.7379


     FLUX ACROSS PARENT-CHILD INTERFACE AT TIME STEP  1 IN STRESS PERIOD     1
     ----------------------------------------------------------------------------


        CUMULATIVE VOLUMES      L**3       RATES FOR THIS TIME STEP      L**3/T
        ------------------                 -----------------------


     TOTAL IN TO CHILD =   2857966.7986     TOTAL IN TO CHILD =      2857.9668

  TOTAL OUT TO PARENT =        0.0000   TOTAL OUT TO PARENT =          0.0000
```

SEAWATER-LGR

This sample problem is derived from the SEAWATER sample problem described in the original GWM manual (Ahlfeld and others, 2005). Seawater intrusion is to be controlled in a coastal area where groundwater is to be pumped as part of a water-supply system (fig. 8). The problem demonstrates the use of decision variables and constraints defined for both parent and child models, linear-summation constraints that include decision variables defined for both models, and the SLP solution method for a nonlinear management formulation. The formulation is nonlinear because the aquifer is modeled as unconfined.

The groundwater-flow system consists of an upper unconfined aquifer with a uniform bottom elevation of 10 ft below local sea level and a lower confined aquifer with a bottom elevation of 20 ft below local sea level. The flow system extends 6,000 ft landward from the coast to a mountainous area underlain by impermeable rocks. The width of the modeled area parallel to the coast is 4,000 ft. Groundwater flow is steady state, and a single stress period of 1 day is simulated.

The parent model consists of 20 rows and 30 columns, and each grid cell is 200 ft by 200 ft (fig. 8). The parent grid consists of two model layers separated by a confining unit with a very low hydraulic conductivity. A single child model with a horizontal grid refinement of 3:1 is located near the coast (fig. 8). The child model is not refined in the vertical dimension, and therefore its layers are coincident with those of the parent grid. The hydraulic properties of each aquifer are homogeneous and isotropic and the same for parent and child models. The hydraulic conductivity of the unconfined aquifer is 5 ft/d; the transmissivity of the confined aquifer is 800 ft^2/d. The confining unit is modeled implicitly by use of a vertical conductance between the layers equal to 0.05 d^{-1}.

Seawater at the coast is modeled by constant heads in column 30 of both layers of the parent model. No-flow boundaries are used along the northern, western, and southern boundaries of each layer. The upper aquifer is recharged at a uniform rate of 0.002 ft/d. The recharge is specified with the RCH Package for both parent and child models and with a **ZONE** file for the parent model to distribute recharge at the interface cells between the parent and child grids. Unlike the original SEAWATER sample problem, there is no unmanaged withdrawal from the aquifer.

The MODFLOW input files consist of **NAME**, **DIS**, **BAS6**, and **RCH** files for each model (and a **ZONE** file for the parent model) and common **BCF6** and **GMG** files used by both models. A head-change closure criterion of 1.0×10^{-8} ft is set in the **GMG** file. LGR closure criteria of 1.0×10^{-7} ft and 1.0×10^{-7} ft^3/d are set for HCLOSELGR and FCLOSELGR, respectively.

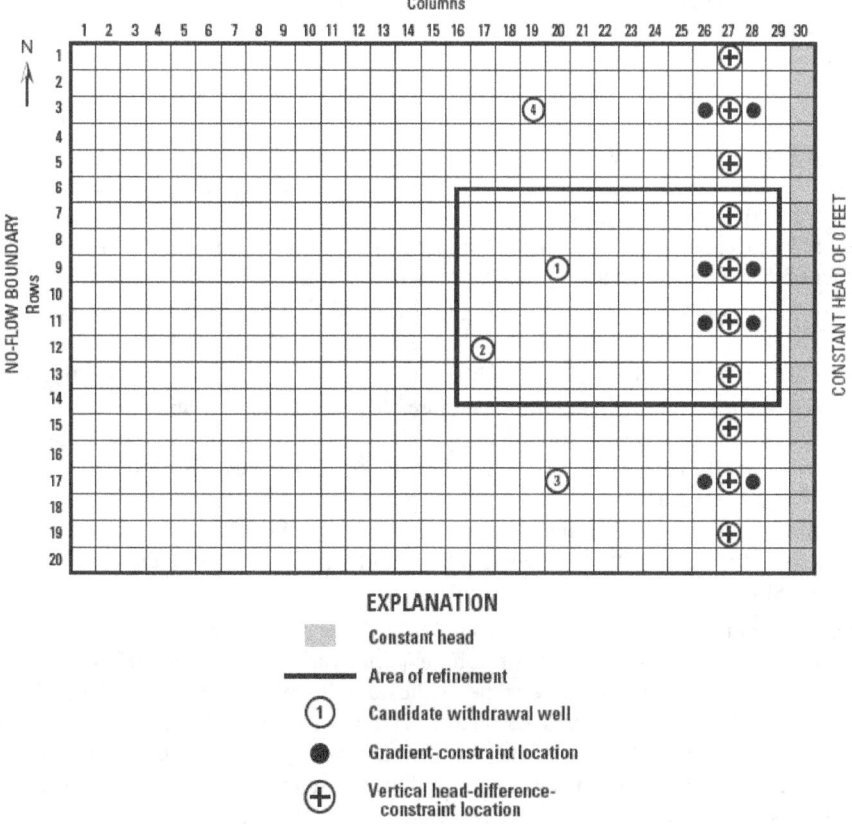

EXPLANATION

- Constant head
- Area of refinement
- (1) Candidate withdrawal well
- ● Gradient-constraint location
- ⊕ Vertical head-difference-constraint location

Figure 8. Schematic diagram showing model grid for SEAWATER-LGR sample problem with local area of refinement.

Mixed-Binary Nonlinear Groundwater-Management Formulation

The water-supply demands of the area are to be met by a combination of groundwater pumped from a maximum of three of four candidate supply wells and surface water imported from a reservoir outside the basin. The surface-water source is represented mathematically by an external variable (variable E, in cubic feet per day). The objective function, which is specified in the **OBJFNC** file for the parent model, is to minimize the use of the imported water (that is, minimize E).

The four candidate wells are designed to pump water equally from the unconfined and confined aquifers. Wells W1 and W2 are within the boundary of the child grid, and wells W3 and W4 are within the parent grid. Binary variables (BV1, BV2, …) are associated with each of the four wells so that a constraint can be implemented that requires that the total number of active wells in both models is less than or equal to three:

$$BV1 + BV2 + BV3 + BV4 \leq 3$$

The constraint is specified in the **SUMCON** file for the parent model. GWM users should be aware that the use of binary variables in nonlinear management problems such as this can cause convergence problems with the SLP solution approach; however, in this case, the same set of wells is selected in each iteration of the SLP solution process, so convergence problems are not of concern. A second linear-summation constraint is defined to require that the total amount of water supplied from the candidate wells and surface-water reservoir is greater than or equal to 50,000 ft³/d:

$$W1 + W2 + W3 + W4 + E \geq 50,000$$

Two types of head constraints are set near the coast to control landward migration of seawater. The first are vertical head-difference constraints set at 10 locations shown on figure 8 to maintain a downward head difference of 0.005 ft between the upper and lower aquifers at each site. The second are horizontal head-gradient constraints set at four locations to maintain a seaward hydraulic gradient of at least 2.80 ft/400 ft (0.007 ft/ft) in each layer of the model (for a total of eight gradient constraints—four in the top layer at each site and four in the bottom layer). The head constraints are specified in **HEDCON** files for both the parent and child models.

Although the objective function and many of the constraints of the problem are linear, the head constraints in the water-table aquifer are nonlinear. Therefore, SLP with binary variables (a mixed-binary nonlinear management formulation) is used to solve the management problem. Most of the solution variables specified in the **SOLN** file for the problem are unchanged from those used in the original SEAWATER problem; however, SLPITMAX is reduced from 10 to 5, LPITMAX is reduced from 1,000 to 100, and BBITMAX is set to 200. Two closure criteria for the SLP algorithm are specified in the **SOLN** file, one for the withdrawal rates at the candidate wells (variable SLPVCRIT, which was set to 1.0 ft³/d) and one for the value of the objective function (variable SLPZCRIT, which was set to 0.001 ft³).

Three iterations of the SLP algorithm are required to meet the SLP closure criteria. The value of the objective function at the solution is 32,124 ft³ of water imported each day from the surface-water reservoir. Nonzero withdrawals are calculated for three of the candidate wells: W1 (5,899 ft³/d), W3 (6,534 ft³/d), and W4 (5,441 ft³/d). Therefore, well W2 is undeveloped. The total water supplied to the system is just equal to 50,000 ft³/d, 32,124 ft³/d from surface water and 17,876 ft³/d from groundwater. Three of the gradient constraints are binding in the solution, two in the top layer of the parent model (gd-1 and gd-7) and one in the top layer of the child model (gd-3).

The identical management problem was solved with a single grid with no refinement to enable comparison of the two solutions. The optimal solution found for the problem is 32,128 ft³/d imported from the surface-water reservoir. Nonzero withdrawals were again calculated for the same three wells: W1 (5,885 ft³/d), W3 (6,538 ft³/d), and W4 (5,448 ft³/d). The differences in the solutions between the two simulation approaches are the result of small differences in model-calculated heads at each GWM perturbation; these head differences resulted in different response matrices for each solution.

Sensitivity Analysis of Convergence Criteria

As described above, the SLP algorithm requires that two convergence (or closure) criteria—SLPVCRIT and SLPZCRIT—be specified for solution of the optimization problem. These criteria are in addition to the convergence criterion that must be specified for the MODFLOW GMG matrix solver (HCLOSE) and the two closure criteria that must be specified for the LGR algorithm (HCLOSELGR and FCLOSELGR). Therefore, three levels of convergence need to be appropriately assigned to satisfactorily solve the optimization problem. As noted earlier in the background description of LGR, Mehl and Hill (2005) recommend that the convergence criterion used for the matrix solver be less than or equal to that used for the LGR head criterion (HCLOSELGR) so that the accuracy of the head values for an individual model are more precise than the coupled head-change boundary condition for the child model(s).

A sensitivity analysis of the GMG and LGR closure criteria was done to illustrate their effects on the value of the objective function for the sample problem. HCLOSE was varied in a series of simulations over a range of 1.0×10^{-8} to 1.0×10^{-3} ft, while HCLOSELGR and FCLOSELGR were simultaneously set two orders of magnitude greater than HCLOSE in each simulation. The SLP criteria were held constant in all simulations (1.0 ft^3/d for SLPVCRIT and 0.001 ft^3 for SLPZCRIT). The results of the simulations are summarized in table 3.

The solution to the nonlinear SLP problem begins to degrade slightly as the GMG and LGR closure criteria are relaxed (table 3). The maximum discrepancy in the value of the objective function is only 0.0011 percent when the value of HCLOSE is varied from 1.0×10^{-8} and 1.0×10^{-4} ft, and the SLP algorithm converged in three iterations in each simulation. However, when HCLOSE is increased to a value of 1.0×10^{-3} ft, the precision of the response coefficients calculated by GWM is inadequate, and the SLP algorithm fails to converge on its third iteration. Also, with HCLOSE held at 1.0×10^{-8} ft and HCLOSELGR and FCLOSELGR equal to 1.0×10^{0}, four iterations of the SLP algorithm are required (see last row of table 3), in contrast to three iterations for the previous runs. This run demonstrates that relatively large values of HCLOSELGR and FCLOSELGR can be used for this problem without much reduction in the accuracy of the value of the objective function.

The SLP convergence criteria are independent of the closure criteria of both the GMG matrix solver and the LGR parent-child model iterations. The values of SLPVCRIT and SLPZCRIT are the maximum changes in the values of the withdrawal/injection rates and objective function, respectively. Therefore, their relation to the head-change criteria will depend on the particular aquifer system being simulated. Generally, the SLP closure criteria should be small enough to produce no appreciable change in the optimal solution while simultaneously limiting the number of iterations required for a GWM solution. Before closure criteria for the SLP algorithm are selected, appropriate closure criteria for the matrix and LGR solvers that result in accurate heads on the parent and child models should be determined.

Table 3. Dependence of the value of the objective function of the SEAWATER-LGR sample problem on MODFLOW and LGR closure criteria.

[HCLOSELGR is measured in feet, and FCLOSELGR in cubic feet per day]

HCLOSE for MODFLOW GMG solver (feet)	HCLOSELGR and FCLOSELGR for LGR	Value of the objective function (cubic feet)
1.0×10^{-8}	1.0×10^{-6}	32,124.42
1.0×10^{-6}	1.0×10^{-4}	32,124.42
1.0×10^{-4}	1.0×10^{-2}	32,124.06
1.0×10^{-3}	1.0×10^{-1}	Precision of response coefficient inadequate
1.0×10^{-8}	1.0×10^{0}	32,036.31

Selected Input and Output Files for Sample Problem

LGR CONTROL file (*sealgr.lgr*)

```
LGR                     ; LGR keyword
2                       ;NGRIDS
..\data\SEAWATER-LGR\sealgr_par.nam          ;NAMEFILE for parent
PARENTONLY              ;GRIDSTATUS
00 00                   ;IUPBHSV,IUPBFSV
..\data\SEAWATER-LGR\sealgr_chd.nam          ;NAMEFILE for child
CHILDONLY               ;GRIDSTATUS
1 -59  0  0             ;ISHFLG, IBFLG, IUCBHSV, IUCBFSV
50 0                    ;MXLGRITER, IOUTLGR
0.500 0.500             ;RELAXH, RELAXF     relaxation for heads and fluxes
1.0E-7 1.0E-7           ;HCLOSELGR, FCLOSELGR
1 6  16                 ;NPLBEG,NPRBEG,NPCBEG
2 14 29                 ;NPLEND,NPREND,NPCEND
3                       ;NCPP  of child cells per width of parent
1 1                     ;NCPPL  # of child layers per parent layer
```

NAME file for parent model (*sealgr_par.nam*)

```
LIST   10  sealgr_par.lst
DIS    11  ..\data\SEAWATER-LGR\sealgr_par.dis
BAS6   12  ..\data\SEAWATER-LGR\sealgr_par.ba6
BCF6   13  ..\data\SEAWATER-LGR\sealgr.bc6
RCH    15  ..\data\SEAWATER-LGR\sealgr_par.rch
GMG    16  ..\data\SEAWATER-LGR\sealgr.gmg
ZONE   17  ..\data\SEAWATER-LGR\sealgr_par.zone
GWM    18  ..\data\SEAWATER-LGR\sealgr_par.gwm
```

GWM file for parent model (*sealgr_par.gwm*)

```
#SEAWATER-LGR Sample Problem, GWM Parent file
OUT      sealgr_par.out
DECVAR   ..\data\SEAWATER-LGR\sealgr_par.decvar
OBJFNC   ..\data\SEAWATER-LGR\sealgr_par.objfnc
VARCON   ..\data\SEAWATER-LGR\sealgr_par.varcon
SUMCON   ..\data\SEAWATER-LGR\sealgr_par.sumcon
HEDCON   ..\data\SEAWATER-LGR\sealgr_par.hedcon
SOLN     ..\data\SEAWATER-LGR\sealgr_par.soln
```

Decision variable (DECVAR) file specified for parent model (*sealgr_par.decvar*)

```
#SEAWATER-LGR Sample Problem, DECVAR Parent file
 1 0                         #1-IPRN  GWMWFILE
 2 1 2                       #2-NFVAR  NEVAR  NBVAR
W3  2 0 0 0 W  Y  1    #3a-FVNAME NC LAY ROW COL FTYPE FSTAT WSP
     0.500 1 17 20
     0.500 2 17 20
W4  2 0 0 0 W  Y  1
     0.500 1  3 19
     0.500 2  3 19
E   IM  1                    #EVNAME  ETYPE  ESP
BV3  1  W3                   #BVNAME  NDV  BVLIST
BV4  1  W4
```

Decision-variable constraints (VARCON) file specified for parent model (*sealgr_par.varcon*)

```
#SEAWATER-LGR Sample Problem, VARCON Parent file
 1                       #1-IPRN
W3 0.0d0  1.0d4  0.0d0   #2-FVNAME FVMIN FVMAX FVREF
W4 0.0d0  1.0d4  0.0d0
E  0.0d0  1.0d6
```

Objective Function (OBJFNC) file specified for parent model (*sealgr_par.objfnc*)

```
#SEAWATER-LGR Sample Problem, OBJFNC file
1          #1-IPRN
MIN  WSDV  #2-OBJTYP FNTYP
0  1  0    #3-NFVOBJ NEVOBJ NBVOBJ
E  1.00    #4-EVNAME EVOBJC
```

Linear-summation constraint (SUMCON) file for parent model (*sealgr_par.sumcon*)

```
#SEAWATER-LGR Sample Problem, SUMCON file
1                    #1-IPRN
2                    #2-SMCNUM
demand 5 ge 5.0d4    #3a-SMCNAME NTERMS TYPE RHS
 W1  1.0             #3b-GVNAME GVCOEFF
 W2  1.0
 W3  1.0
 W4  1.0
  E  1.0
wellson 4 le 3       #3a-SMCNAME NTERMS TYPE RHS
 BV4  1.0            #3b-GVNAME GVCOEFF
 BV3  1.0
 BV2  1.0
 BV1  1.0
```

Head constraints (HEDCON) file for parent model (*sealgr_par.hedcon*)

```
#SEAWATER-LGR Sample Problem, HEDCON Parent file
1                        #1-IPRN
0  0  6  4               #2-NHB NDD NDF NGD
hd-01 1 1 27   2 1 27 0.005 1    #5-HDIFNAME LAY1 ROW1 COL1 LAY2 ROW2 COL2 HD NSP
hd-02 1 3 27   2 3 27 0.005 1
hd-03 1 5 27   2 5 27 0.005 1
hd-08 1 15 27  2 15 27 0.005 1
hd-09 1 17 27  2 17 27 0.005 1
hd-10 1 19 27  2 19 27 0.005 1
gd-1 1 3 26 1 3 28 400. 0.007 1 #6-GRADNAME (L,R,C)1 (L,R,C)2 LEN GRAD NSP
gd-2 2 3 26 2 3 28 400. 0.007 1
gd-7 1 17 26 1 17 28 400. 0.007 1
gd-8 2 17 26 2 17 28 400. 0.007 1
```

NAME file for child model (*sealgr_chd.nam*)

```
LIST  20  sealgr_chd.lst
DIS   21  ..\data\SEAWATER-LGR\sealgr_chd.dis
BAS6  22  ..\data\SEAWATER-LGR\sealgr_chd.ba6
BCF6  23  ..\data\SEAWATER-LGR\sealgr.bc6
RCH   25  ..\data\SEAWATER-LGR\sealgr_chd.rch
GMG   26  ..\data\SEAWATER-LGR\sealgr.gmg
GWM   28  ..\data\SEAWATER-LGR\sealgr_chd.gwm
```

GWM file for child model (*sealgr_chd.gwm*)

```
#SEAWATER-LGR Sample Problem, GWM Child file
DECVAR  ..\data\SEAWATER-LGR\sealgr_chd.decvar
VARCON  ..\data\SEAWATER-LGR\sealgr_chd.varcon
HEDCON  ..\data\SEAWATER-LGR\sealgr_chd.hedcon
```

Decision variable (DECVAR) file specified for child model (*sealgr_chd.decvar*)

```
#SEAWATER-LGR Sample Problem, DECVAR Child file
1 0                      #1-IPRN  GWMWFILE
2 0 2                    #2-NFVAR  NEVAR  NBVAR
W1 2 0 0 0  W  Y  1      #3a-FVNAME NC LAY ROW COL FTYPE FSTAT WSP
    0.500  1  10 13
    0.500  2  10 13
W2 2 0 0 0  W  Y  1
    0.500  1  19  4
    0.500  2  19  4
BV1  1  W1               #BVNAME  NDV  BVLIST
BV2  1  W2
```

Decision-variable constraints (VARCON) file specified for child model (*sealgr_chd.varcon*)

```
#SEAWATER-LGR Sample Problem, VARCON Child file
1                        #1-IPRN
W1 0.0d0  1.0d4  0.0d0    #2-FVNAME FVMIN FVMAX FVREF
W2 0.0d0  1.0d4  0.0d0
```

Part of the GWM OUT file (*sealgr_par.out*)

```
-----------------------------------------------------------------------
               Ground-Water Management Solution
-----------------------------------------------------------------------


       OPTIMAL SOLUTION FOUND

       OPTIMAL RATES FOR EACH FLOW VARIABLE
       ----------------------------------------

Variable          Withdrawal          Injection          Contribution
Name              Rate                Rate               To Objective
----------        --------------      ------------       ------------
  W3              6.534729E+03                           0.000000E+00
  W4              5.441153E+03                           0.000000E+00
  W1              5.899694E+03                           0.000000E+00
  W2              0.000000E+00                           0.000000E+00
                  --------------      ------------       ------------
TOTALS            1.787558E+04        0.000000E+00       0.000000E+00

       OPTIMAL RATES FOR EACH EXTERNAL VARIABLE
       ----------------------------------------

Variable          Export              Import             Contribution
Name              Rate                Rate               To Objective
----------        --------------      ------------       ------------
  E                                   3.212442E+04       3.212442E+04
                  --------------      ------------       ------------
TOTALS            0.000000E+00        3.212442E+04       3.212442E+04

       OPTIMAL VALUES FOR EACH BINARY VARIABLE
       ----------------------------------------

Variable                                                 Contribution
Name              Value                                  To Objective
----------        ------------                           ------------
  BV3             1                                      0.000000E+00
  BV4             1                                      0.000000E+00
  BV1             1                                      0.000000E+00
  BV2             0                                      0.000000E+00
                  ------------                           ------------
TOTALS            3                                      0.000000E+00


       OBJECTIVE FUNCTION VALUE                          3.212442E+04


       BINDING CONSTRAINTS
Constraint Type       Name      Status      Shadow Price
---------------       ----      ------      ------------
Head Gradient         gd-1      Binding     2.3181E+03
Head Gradient         gd-7      Binding     3.3428E+03
Head Gradient         gd-3      Binding     3.2667E+03
Summation             demand    Binding     1.0000E+00
Summation             wellson   Binding     Not Available
Minimum Flow Rate     W2        Binding     Not Available
```

SUPPLY-3GRID

This sample problem demonstrates the use of multiple areas of refinement in a steady-state regional water-supply problem. The problem highlights the use of binary variables associated with decision variables on different grids, summation constraints with decision variables on different grids, and head constraints on a single grid.

The hypothetical confined aquifer is 50,000 ft long, 30,000 ft wide, and 200 ft thick. The parent model consists of 30 rows, 50 columns, and a single layer (fig. 9). Each grid cell is 1,000 ft by 1,000 ft in the horizontal dimension. A single, steady-state stress period of 1,000 days is used to simulate flow in the system. The aquifer is heterogeneous and isotropic. The transmissivity in the western half of the aquifer is 8,000 ft²/d and 2,000 ft²/d in the eastern half. The aquifer is bounded on the north and south by no-flow boundary conditions and on the east and west by specified-head boundary conditions. Specified heads of 300 ft on the eastern boundary and 150 ft on the western boundary result in a hydraulic gradient of 0.003 ft/ft in the absence of withdrawals.

Two supply systems are being considered for development. The first is on the western side of the aquifer and consists of three candidate wells, Q1, Q2, and Q3; the second is on the eastern side of the aquifer and consists of two candidate wells, Q4 and Q5 (fig. 9). Three child models are used to simulate flow conditions near the supply systems, and each grid uses a 5:1 horizontal-refinement ratio and 1:1 vertical-refinement ratio.

Mixed-Binary Linear Groundwater-Management Formulation

As in the previous sample problem, water is available from a surface-water reservoir that is external to the basin (represented mathematically as variable E, in cubic feet per day), and the management objective is to minimize the use of this external source (that is, minimize E). Each well has a maximum withdrawal rate of 500,000 ft³/d; the external source has a maximum supply rate of 1.0×10^6 ft³/d. The external variable could be defined in the **DECVAR** file for either the parent or any one of the child models; here it is specified with the westernmost child model (child model 1).

Two linear-summation constraints are specified in the **SUMCON** file for the parent model. The first requires that the sum of the withdrawals from the five candidate wells and from the external surface-water source meets a minimum demand of 1.0×10^6 ft³/d:

$$Q1 + Q2 + Q3 + Q4 + Q5 + E \geq 1.0 \times 10^6$$

The second constraint specifies that only one of the supply systems is developed. This is accomplished by introducing two binary variables: variable BV1 is associated with the western system (wells Q1, Q2, and Q3) and BV2 is associated with the eastern system (wells Q4 and Q5). The binary variables could be defined in either the parent model or any one of the three child models; here they are defined in the child model with well Q5 (child model 3). The summation constraint is written

$$BV1 + BV2 \leq 1.0.$$

Four head constraints are imposed in the parent model to ensure that heads remain greater than or equal to 210 ft at the four locations (fig. 9).

The management formulation is a mixed-binary linear formulation because binary variables are used, the aquifer is confined, and the objective function and all of the constraints are linear. Therefore, the LP solution option of GWM was used with appropriate values specified for the solution parameters associated with binary variables (that is, BBITMAX and BBITPRT).

The optimal solution is to develop the western side of the aquifer in the vicinity of wells Q1, Q2, and Q3; however, only wells Q1 and Q3 are developed, with their withdrawal rates equal to 500,000 ft³/d for Q1 (its upper bound) and 425,693 ft³/d for Q3. The value of the objective function is 7.4307×10^7 ft³, which is equal to the optimal rate of supply from the surface-water reservoir (74,307 ft³/d) multiplied by 1,000 days. Head constraint hd-1, which is closest to the two pumping wells (row 14, column 24 of the parent model), is binding. Selected input files for the problem are provided below.

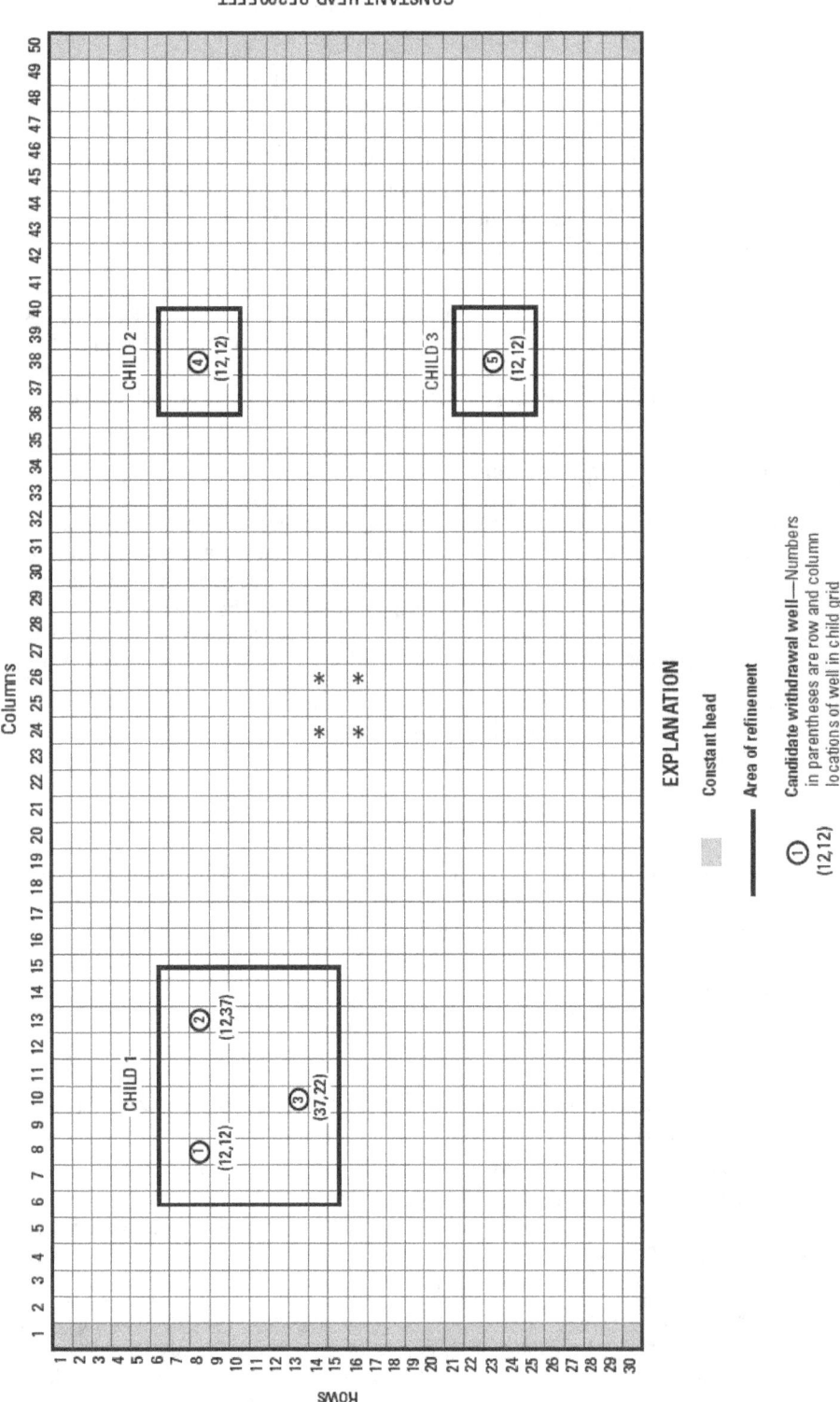

Figure 9. Schematic diagram showing model grid for SUPPLY-3GRID sample problem with three areas of local grid refinement.

Selected Input Files for Sample Problem

LGR CONTROL file (*sup3g.lgr*)

```
LGR                      ;LGR keyword
4                        ;NGRIDS
..\data\SUPPLY-3GRID\sup3g_par.nam        ;NAMEFILE for parent
PARENTONLY               ;GRIDSTATUS
 0  0                    ;IUPBFSV, IUPBHSV
..\data\SUPPLY-3GRID\sup3g_c1.nam         ;NAMEFILE for child 1
CHILDONLY                ;GRIDSTATUS
1 -59  0  0              ;ISHFLG, IBFLG, IUCBHSV, IUCBFSV
50 0                     ;MXLGRITER, IOUTLGR
0.50 0.50                ;RELAXH, RELAXF   relaxation for heads and fluxes
1.0E-4 1.0E-4            ;HCLOSELGR, FCLOSELGR
1  6  6                  ;NPLBEG,NPRBEG,NPCBEG
1 15 15                  ;NPLEND,NPREND,NPCEND
5                        ;NCPP
1                        ;NCPPL
..\data\SUPPLY-3GRID\sup3g_c2.nam         ;NAMEFILE for child 2
CHILDONLY                ;GRIDSTATUS
1 -58  0  0              ;ISHFLG, IBFLG, IUCBHSV, IUCBFSV
50 0                     ;MXLGRITER, IOUTLGR
0.50 0.50                ;RELAXH, RELAXF   relaxation for heads and fluxes
1.0E-4 1.0E-4            ;HCLOSELGR, FCLOSELGR
1  6 36                  ;NPLBEG,NPRBEG,NPCBEG
1 10 40                  ;NPLEND,NPREND,NPCEND
5                        ;NCPP
1                        ;NCPPL
..\data\SUPPLY-3GRID\sup3g_c3.nam         ;NAMEFILE for child 3
CHILDONLY                ;GRIDSTATUS
1 -57  0  0              ;ISHFLG, IBFLG, IUCBHSV, IUCBFSV
50 0                     ;MXLGRITER, IOUTLGR
0.50 0.50                ;RELAXH, RELAXF   relaxation for heads and fluxes
1.0E-4 1.0E-4            ;HCLOSELGR, FCLOSELGR
1 21 36                  ;NPLBEG,NPRBEG,NPCBEG
1 25 40                  ;NPLEND,NPREND,NPCEND
5                        ;NCPP
1                        ;NCPPL
```

NAME file for parent model (*sup3g_par.nam*)

```
LIST   10 sup3g_par.lst
BAS6   11 ..\data\SUPPLY-3GRID\sup3g_par.ba6
BCF6   66 ..\data\SUPPLY-3GRID\sup3g_par.bc6
DIS    13 ..\data\SUPPLY-3GRID\sup3g_par.dis
PCG    14 ..\data\SUPPLY-3GRID\sup3g_par.pcg
ZONE   15 ..\data\SUPPLY-3GRID\sup3g_par.zone
RCH    16 ..\data\SUPPLY-3GRID\sup3g_par.rch
GWM    17 ..\data\SUPPLY-3GRID\sup3g_par.gwm
```

GWM file for parent model (*sup3g_par.gwm*)

```
#SUP3G Sample Problem, parent GWM file
OUT        sup3g_par.out
OBJFNC     ..\data\SUPPLY-3GRID\sup3g_par.objfnc
SUMCON     ..\data\SUPPLY-3GRID\sup3g_par.sumcon
HEDCON     ..\data\SUPPLY-3GRID\sup3g_par.hedcon
SOLN       ..\data\SUPPLY-3GRID\sup3g_par.soln
```

Objective Function (OBJFNC) file specified for parent model (*sup3g_par.objfnc*)

```
#SUP3G Sample Problem, OBJFNC file
1           #1-IPRN
MIN  WSDV       #2-OBJTYP FNTYP
0  1  0         #3-NFVOBJ NEVOBJ  NBVOBJ
E    1          #5-EVNAME  EVOBJC
```

Linear-summation constraint (SUMCON) file for parent model (*sup3g_par.sumcon*)

```
#SUP3G Sample Problem, SUMCON file (sup3g_par.sumcon)
1                       #1-IPRN
2                       #2-SMCNUM
DEMAND 6 GE 1e6         #3a-SMCNAME NTERMS TYPE RHS
 Q1  1.0                #3b-GVNAME GVCOEFF
 Q2  1.0
 Q3  1.0
 Q4  1.0
 Q5  1.0
  E  1.0
FIELDS 2 LE 1
 BV1  1
 BV2  1
```

Head constraints (HEDCON) file for parent model (*sup3g_par.hedcon*)

```
#SUP3G Sample Problem, HEDCON file
1                           #1-IPRN
4  0  0  0                  #2-NHB NDD NDF NGD
hd-1  1  14  24  ge  210.0  1    #3-HBNAME LAYH ROWH COLH TYPH BND NSP
hd-2  1  14  26  ge  210.0  1
hd-3  1  16  24  ge  210.0  1
hd-4  1  16  26  ge  210.0  1
```

NAME file for child model 1 (*sup3g_c1.nam*)

```
LIST    20 sup3g_c1.lst
BAS6    21 ..\data\SUPPLY-3GRID\sup3g_c1.ba6
BCF6    22 ..\data\SUPPLY-3GRID\sup3g_c1.bc6
DIS     23 ..\data\SUPPLY-3GRID\sup3g_c1.dis
PCG     24 ..\data\SUPPLY-3GRID\sup3g_chd.pcg
RCH     25 ..\data\SUPPLY-3GRID\sup3g_chd.rch
GWM     26 ..\data\SUPPLY-3GRID\sup3g_c1.gwm
```

GWM file for child model 1 (*sup3g_c1.gwm*)

```
#SUP3G Sample Problem, child #1 GWM file
DECVAR      ..\data\SUPPLY-3GRID\sup3g_c1.decvar
VARCON      ..\data\SUPPLY-3GRID\sup3g_c1.varcon
```

Decision variable (DECVAR) file specified for child model 1 (*sup3g_c1.decvar*)

```
#SUP3G Sample Problem, child #1 DECVAR file
 1 0                        #1-IPRN  GWMWFILE
 3 1 0                      #2-NFVAR NEVAR NBVAR
Q1  1  1  12  12  W  Y  1   #3a-FVNAME NC LAY ROW COL FTYPE FSTAT WSP
Q2  1  1  12  37  W  Y  1
Q3  1  1  37  22  W  Y  1
E    IM    1                #4-EVNAME ETYPE ESP
```

Decision-variable constraints (VARCON) file specified for child model 1 (*sup3g_c1.varcon*)

```
#SUP3G Sample Problem, child #1 VARCON file
1                       #1-IPRN
Q1 0.0d0  5.0d5  0.0d0   #2-FVNAME FVMIN FVMAX FVREF
Q2 0.0d0  5.0d5  0.0d0
Q3 0.0d0  5.0d5  0.0d0
E  0.0d0  1.0d6
```

NAME file for child model 2 (*sup3g_c2.nam*)

```
LIST   30 sup3g_c2.lst
BAS6   31 ..\data\SUPPLY-3GRID\sup3g_c2.ba6
BCF6   32 ..\data\SUPPLY-3GRID\sup3g_c2.bc6
DIS    33 ..\data\SUPPLY-3GRID\sup3g_c2.dis
PCG    34 ..\data\SUPPLY-3GRID\sup3g_chd.pcg
RCH    35 ..\data\SUPPLY-3GRID\sup3g_chd.rch
GWM    36 ..\data\SUPPLY-3GRID\sup3g_c2.gwm
```

GWM file for child model 2 (*sup3g_c2.gwm*)

```
#SUP3G Sample Problem, child #2 GWM file
DECVAR      ..\data\SUPPLY-3GRID\sup3g_c2.decvar
VARCON      ..\data\SUPPLY-3GRID\sup3g_c2.varcon
```

Decision variable (DECVAR) file specified for child model 2 (*sup3g_c2.decvar*)

```
#SUP3G Sample Problem, child #2 DECVAR file
 1 0                          #1-IPRN  GWMWFILE
 1 0 0                        #2-NFVAR NEVAR NBVAR
 Q4  1  1  12  12  W  Y  1    #3a-FVNAME NC LAY ROW COL FTYPE FSTAT WSP
```

Decision-variable constraints (VARCON) file specified for child model 2 (*sup3g_c2.varcon*)

```
#SUP3G Sample Problem, child #2 VARCON file
 1                            #1-IPRN
 Q4 0.0d0  5.0d5  0.0d0       #2-FVNAME FVMIN FVMAX FVREF
```

NAME file for child model 3 (*sup3g_c3.nam*)

```
LIST   40 sup3g_c3.lst
BAS6   41 ..\data\SUPPLY-3GRID\sup3g_c3.ba6
BCF6   42 ..\data\SUPPLY-3GRID\sup3g_c3.bc6
DIS    43 ..\data\SUPPLY-3GRID\sup3g_c3.dis
PCG    44 ..\data\SUPPLY-3GRID\sup3g_chd.pcg
RCH    45 ..\data\SUPPLY-3GRID\sup3g_chd.rch
GWM    46 ..\data\SUPPLY-3GRID\sup3g_c3.gwm
```

GWM file for child model 3 (*sup3g_c3.gwm*)

```
#SUP3G Sample Problem, child #3 GWM file
DECVAR      ..\data\SUPPLY-3GRID\sup3g_c3.decvar
VARCON      ..\data\SUPPLY-3GRID\sup3g_c3.varcon
```

Decision variable (DECVAR) file specified for child model 3 (*sup3g_c3.decvar*)

```
#SUP3G Sample Problem, child #3 DECVAR file
 1 0                          #1-IPRN  GWMWFILE
 1 0 2                        #2-NFVAR NEVAR NBVAR
 Q5  1  1  12  12  W  Y  1    #3a-FVNAME NC LAY ROW COL FTYPE FSTAT WSP
 BV1  3  Q1  Q2  Q3           #BVNAME  NDV  BVLIST
 BV2  2  Q4  Q5               #BVNAME  NDV  BVLIST
```

Decision-variable constraints (VARCON) file specified for child model 3 (*sup3g_c3.varcon*)

```
#SUP3G Sample Problem, child #3 VARCON file
 1                            #1-IPRN
 Q5 0.0d0  5.0d5  0.0d0       #2-FVNAME FVMIN FVMAX FVREF
```

Acknowledgments

The authors thank Steffen Mehl, California State University (Chico) and USGS, for his advice and assistance during the development of the LGR capability for GWM-2005 and for his review of a draft of this report. The authors also thank Edward Banta and Mary Ashman of the USGS for their helpful review comments on the draft report.

References Cited

Ahlfeld, D.P., Barlow, P.M., and Mulligan, A.E., 2005, GWM—A Ground-Water Management Process for the U.S. Geological Survey modular ground-water model (MODFLOW-2000): U.S. Geological Survey Open-File Report 2005–1072, 124 p.

Ahlfeld, D.P., and Baro-Montes, Gemma, 2008, Solving unconfined groundwater flow management problems with successive linear programming: Journal of Water Resources Planning and Management, v. 134, no. 5, p. 404–412.

Ahlfeld, D.P., and Mulligan, A.E., 2000, Optimal management of flow in groundwater systems: San Diego, CA, Academic Press, 185 p.

Dantzig, G.B., 1963, Linear programming and extensions: Princeton, NJ, Princeton University Press, 627 p.

Gass, Saul, 1985, Linear programming—Methods and applications: New York, McGraw-Hill, 532 p.

Halford, K.J., and Hanson, R.T., 2002, User guide for the drawdown-limited, multi-node well (MNW) Package for the U.S. Geological Survey's modular three-dimensional finite-difference ground-water flow model, versions MODFLOW-96 and MODFLOW-2000: U.S. Geological Survey Open-File Report 02–293, 33 p.

Harbaugh, A.W., 1995, Direct solution package based on alternating diagonal ordering for the U.S. Geological Survey modular finite-difference ground-water flow model: U.S. Geological Survey Open-File Report 95–288, 46 p.

Harbaugh, A.W., 2005, MODFLOW-2005, The U.S. Geological Survey modular ground-water model—The Ground-Water Flow Process: U.S. Geological Survey Techniques and Methods 6–A16 [variously paged].

Harbaugh, A.W., Banta, E.R., Hill, M.C., and McDonald, M.G., 2000, MODFLOW-2000, the U.S. Geological Survey modular ground-water model—User guide to modularization concepts and the Ground-Water Flow Process: U.S. Geological Survey Open-File Report 00–92, 121 p.

Mehl, S.W., and Hill, M.C, 2005, MODFLOW-2005, The U.S. Geological Survey modular ground-water model—Documentation of shared node local grid refinement (LGR) and the boundary flow and head (BFH) Package: U.S. Geological Survey Techniques and Methods 6–A16, 68 p.

Mehl, S.W., and Hill, M.C., 2007, MODFLOW-2005, The U.S. Geological Survey modular ground-water model—Documentation of the multiple-refined-areas capability of local grid refinement (LGR) and the boundary flow and head (BFH) package: U.S. Geological Survey Techniques and Methods 6–A21, 13 p.

Nemhauser, G.L., and Wolsey, L.A., 1988, Integer and combinatorial optimization: New York, Wiley, 763 p.

Niswonger, R.G., and Prudic, D.E., 2005, Documentation of the Streamflow-Routing (SFR2) Package to include unsaturated flow beneath streams—A modification to SFR1: U.S. Geological Survey Techniques and Methods 6–A13, 48 p.

Prudic, D.E., 1989, Documentation of a computer program to simulate stream-aquifer relations using a modular, finite-difference, ground-water flow model: U.S. Geological Survey Open-File Report 88–729, 113 p.

Prudic, D.E., Konikow, L.F., and Banta, E.R., 2004, A new Streamflow-Routing (SFR1) Package to simulate stream-aquifer interaction with MODFLOW-2000: U.S. Geological Survey Open-File Report 2004–1042, 95 p.

Pulido-Velazquez, David, Ahlfeld, David, Andreu, Joaquin, and Sahuquillo, Andres, 2008, Reducing the computational cost of unconfined groundwater flow in conjunctive-use models at basin scale assuming linear behaviour—the case of Adra-Campo de Dalias: Journal of Hydrology, v. 353, p. 159–174.

Wilson, J.D., and Naff, R.L., 2004, MODFLOW-2000, the U.S. Geological Survey modular ground-water model—GMG linear equation solver package documentation: U.S. Geological Survey Open-File Report 2004–1261, 47 p.

Appendix

Program Modifications to Convert GWM-2000 to GWM-2005

Contents

Introduction

This appendix describes programming issues related to the conversion of GWM-2000 to GWM-2005. The GWM Process consists of a MODFLOW Basic (BAS) Package, the GWM Response Matrix Solution (RMS) Package for solving the optimization problem, and six other GWM packages for defining decision variables, an objective function, and constraints. To couple the GWM and MODFLOW Groundwater Flow (GWF) Processes, changes to the MAIN program of MODFLOW were required.

Overview of Modifications to GWM-2000 to Create GWM-2005

A major programming change made in the conversion of MODFLOW-2000 to MODFLOW-2005 was the use of the Fortran-90 module structure for storing and sharing data. The components of the GWM Process in GWM-2000 were originally coded according to this data structure so that the incorporation of GWM into MODFLOW-2005 would be fairly straightforward. Nevertheless, minor changes have been made to most of the GWM packages to take advantage of the full use of module structures by MODFLOW-2005. These changes made the new GWM packages incompatible with MODFLOW-2000. As a result, a new set of GWM packages were defined, each with the number designation 2; for example, the package GWM1OBJ1 became GWM1OBJ2. A second set of changes to accommodate the LGR capability led to a new set of GWM packages with number designation 3. To prevent confusion, this change of number designation was made to all GWM packages, even if no internal changes were made to the code. In each case, the version date listed at the beginning of each package has been changed; however, the version date in each subroutine has been unchanged from that in version 1 unless the code within the subroutine has been changed. Major changes have been made to the GWM BAS Package and minor changes to all other packages.

To accommodate the new features of MODFLOW-2005, a new version of the BAS package for GWM was created. This version is named GWM1BAS3 and includes modified versions of the Allocate and Read (AR) and Rewind (RW) procedures, eliminates the RPP subroutine, and introduces three new subroutines and one function. The changes to the AR and RW procedures and the functions of the new subroutines are described here.

Subroutine GWM1BAS3AR allocates and reads all information needed by GWM. Changes from GWM1BAS1AR include the following:

- Accommodations for the elimination of the **GLOBAL** file from MODFLOW-2005.

- The addition of code to look for OUT as the first keyword in the **GWM** file. If present, a name is read and assigned to the GWM output file. If not present, this output goes to a file with the default name GWM.OUT.

- The addition of calls to IGETUNIT to obtain a unit number for the GWM.OUT file. IGETUNIT was available in MODFLOW-2000 but is not in MODFLOW-2005. A copy of this subroutine is placed in GWM1BAS3SUBS for.

Subroutine GWM1BAS3RW performs rewind operations on all input files. It also rewinds the **LIST** file and writes a new header to this file. It is constructed from the rewind subroutine found in the MODFLOW-2000 Parameter Estimation Process, PES1BAS6RW.

New subroutine GWM1BAS3DABAS7 selectively deallocates memory that had been allocated in the BAS7 package. Memory that is originally allocated in SGWF2BAS7ARDIS, SGWF2BAS7ARMZ, or SGWF2BAS7I is deallocated in GWM1BAS3DABAS7.

New subroutine GWM1BAS3RRF is a modified version of the GWF1BAS7AR routine with several important differences:

- Most of the allocation in GWF1BAS7AR is not repeated because these arrays are not deallocated.

- A new subroutine, GWM1BAS3OPEN, is called to check that all files remain open.

- If the GWF Process simulation is steady state, then the solution from the prior simulation run can be used as an initial estimate of the solution, substantially reducing solution time. For this case, unless it is the last simulation, STRT in the Basic Package file (**BAS6**) is not read, and the HNEW array is not reset.

New subroutine GWM1BAS3OPEN is derived from SGWF2BAS7OPEN and makes sure that all files listed in the MODFLOW-2005 **NAME** file are still open. If any file is not open, it is reopened in this subroutine. This check is needed because some packages (for example, PVAL) close their input files after reading is completed.

Modifications to GWM-2005 to Accommodate LGR Capability

GWM-2005 is structured to allow placement of decision variables and constraints in any of the multiple models that may be present when the LGR capability is used. Components of the management problem are defined in the **GWM** file for each model. GWM-2005 reads the input files to assemble a single optimization problem that spans all models. This is accomplished in the AR subroutines in each of the GWM packages. Once a single optimization problem is created, the remaining GWM tasks operate much as they did in prior versions of GWM. As a result, changes to the code in the RMS Package or beyond the changes in the AR subroutines in other GWM packages are minor.

The key structures to accomplish the assembly of a single optimization problem over multiple models are in GWM1BAS3AR. Here, all filenames in all GWM files are stored in an array, FNAMEOPT, so that the files may be accessed in any order. Once the GWM files have been read, common GWM file types can be processed together.

First, all **DECVAR** files are processed by calling subroutine GWM1DCV3AR. This routine reads over the **DECVAR** files for all models to determine the total number of decision variables and assembles all decision-variable information, including variable name, into a single set of arrays that span all models. A single index array maps the decision-variable number back to the model in which it is defined. When grid-specific decision-variable information is accessed, such as cell location, this index array is used to reference the correct model.

Next the **OBJFNC** file identified in the parent **GWM** file is processed by calling subroutine GWM1OBJ3AR. Because the names of all decision variables for the multiple models have been read and stored, any of them can be included in the **OBJFNC** file. This routine operates with decision-variable names only, so there are no significant changes required to accommodate the LGR capability.

The **VARCON** files for all models are processed next by subroutine GWM1DCC3AR. These files define constraints on the decision variables. The subroutine operates by first opening all **VARCON** files. A loop over all decision variables commences with information for each decision variable read from the appropriate **VARCON** file. The mapping from decision-variable number to proper **VARCON** file is accomplished by using the index array defined in subroutine GWM1DCV3AR.

Subroutine GWM1DCV3AR next calls subroutine GWM1SMC3AR, which processes the **SUMCON** file defined in the **GWM** file for the parent model. Because all decision-variable names have been defined previously, the **SUMCON** file can reference any of these names. Because this routine operates with the names of decision variables, no significant changes are required to accommodate the LGR capability. Subroutines GWM1HDC3AR and GWM1STC3AR operate in a similar fashion. First, the **HEDCON** or **STRMCON** files for each model are opened, and the number of constraints defined for each model are counted. With this information, the total number of constraints of each type over all models is determined, and appropriate storage is allocated. Next, the constraint information is read for each model and stored in a single set of arrays that span all models. Indexing arrays are defined so that a constraint can be mapped back to the model in which it was defined. The indexing arrays are used in subroutines GWM1HDC3OS and GWM1STC3OS to match the constraint to the specific grid and cell for which it is defined.

Modifications to the MODFLOW-2005 Main Program to Create GWM-2005

The major addition to the MODFLOW-2005 MAIN Program required by GWM is the introduction of a looping structure around the commands for executing a GWF Process simulation. At the completion of each GWF Process run and before the next run is initiated, GWM-2005 rereads most input. In preparation for rereading input, GWM-2005 rewinds all input files by using the subroutine GWM1BAS3RW. The space utilized by each of the GWF packages is deallocated by using the appropriate GWF DA routines. In addition, selected portions of GWF BAS data are deallocated by using the subroutine GWM1BAS3DABAS7. The rereading of the input and associated reallocation of memory is accomplished by calling the AR procedures for each package. Only some of the GWF BAS data are reread. This rereading is accomplished in subroutine GWM1BAS3RRF. Included in this subroutine is a call to GWM1BAS3OPEN, which performs a check to ensure that all files in the MODFLOW-2005 **NAME** file remain open.

The steps required to create the GWM-2005 MAIN Program are as follows:

1—Insert USE statements for GWM subroutines with other USE statements
```
C GWM: GWM SUBROUTINES CALLED FROM MF2005 MAIN PROGRAM
      USE GWM1BAS3SUBS, ONLY: GWM1BAS3AR,GWM1BAS3RW,GWM1BAS3RRF,
     &                        GWM1BAS3DABAS7,GWM1BAS3AP
      USE GWM1RMS3SUBS, ONLY: GWM1RMS3PL,GWM1RMS3PP,GWM1RMS3FP,
     &                        GWM1RMS3FM,GWM1RMS3AP,GWM1RMS3OT,
     &                        GWM1RMS3OS
      USE GWM1DCV3, ONLY: GWF2DCV3FM,GWF2DCV3BD
      USE GWM1HDC3, ONLY: GWM1HDC3OS
      USE GWM1STC3, ONLY: GWM1STC3OS,GWM1STC3OS2
      USE GWM1RMS3, ONLY: MFCNVRG
```

2—Replace version definition
```
      PARAMETER (VERSION='GWM-2005 Y.Y XXXX,FROM MF-LGR VZ.Z.Z')
```

3—Insert local GWM variables among declaration statements
```
C GWM: LOCAL VARIABLES CREATED FOR GWM
      LOGICAL FIRSTSIM,LASTSIM,FINISH,GWMCNVRG,GWMACTIVE
      INTEGER IPERT,NPERT
      REAL HCLOSET,DUMHCLOSE
      DOUBLE PRECISION, DIMENSION(1):: DEPVALS  ! DUMMY ARRAY NEEDED FOR GWMPLL
      CHARACTER(LEN=200),ALLOCATABLE::FNAMEV(:)
```

3A—Insert code to create storage and initialize flag
```
      ALLOCATE(FNAMEV(NGRIDS))                 ! GWM: CREATE STORAGE
      GWMACTIVE=.FALSE.                        ! INITIALIZE NO GWM FILES
      DUMHCLOSE = 100000.0                     ! INITIALIZE FOR COMPARISON
```

4—Insert code to save name of NAME file
```
      FNAMEV(IGRID)=FNAME                      ! GWM SAVE MF NAME FILES
```

5—Move calls to solver AR subroutines and OBS2...AR subroutines to immediately after the call to GWF2LGR1AR, and insert this code to find smallest HCLOSE
```
C        FIND SMALLEST HCLOSE VALUE ACROSS GRIDS
         IF(IUNIT(9).GT.0)  DUMHCLOSE = MIN(DUMHCLOSE,HCLOSE)    !SIP
         IF(IUNIT(10).GT.0) DUMHCLOSE = MIN(DUMHCLOSE,HCLOSEDE4) !DE4
         IF(IUNIT(13).GT.0) DUMHCLOSE = MIN(DUMHCLOSE,HCLOSEPCG) !PCG
         IF(IUNIT(42).GT.0) DUMHCLOSE = MIN(DUMHCLOSE,HCLOSEGMG) !GMG
```

5A—Insert code to check for GWM files and close loop

```
        IF(IUNIT(56).GT.0)GWMACTIVE=.TRUE.          ! TRUE IF GWM ON ANY GRID
C END LOOP FOR ALLOCATING AND READING DATA FOR EACH GRID
        ENDDO
```

6—After closing loop, insert code for GWM...AR call

```
C GWM: READ GWM PROCESS INFORMATION
        IF(GWMACTIVE)THEN
          CALL SGWF2BAS7PNT(1)                    ! POINT TO PARENT FOR NPER, PERLEN
          CALL GWM1BAS3AR(IOUT,NPER,PERLEN,DUMHCLOSE,VERSION,MFVNAM,
     &                    NGRIDS)
        ENDIF
```

7—Immediately following the previous change, insert code for controlling GWM loops

```
C GWM: INSERT GWM LOOPING CONTROLS
        FIRSTSIM = .TRUE.
        GWMCNVRG = .FALSE.
        FINISH   = .FALSE.
C
C-----PREPARE ITERATION LOOP CONTROLS
        IF(.NOT.GWMACTIVE)THEN       ! GWM NOT ACTIVE; LOOP FOR ONE PASS
          LASTSIM = .TRUE.
        ELSEIF(GWMACTIVE)THEN         ! GWM ACTIVE; LOOP FOR GWM ITERATIONS
          LASTSIM = .FALSE.
          MFCNVRG = .TRUE.
        ENDIF
C
C-----BEGIN ITERATION LOOP FOR GWM PROCESS
        DO 300 WHILE (.NOT. FINISH)
C
C-------PREPARE PERTURBATION LOOP CONTROLS
        IF(.NOT.GWMACTIVE)THEN       ! GWM NOT ACTIVE; LOOP FOR ONE PASS
          IPERT = 1
          NPERT = 1
        ELSEIF(GWMACTIVE)THEN         ! GWM ACTIVE; SET LOOP CONTROLS
          CALL GWM1RMS3PL(IPERT,NPERT,FIRSTSIM,LASTSIM)
        ENDIF
C
C-------BEGIN PERTURBATION LOOP FOR GWM PROCESS
        DO 200 WHILE (IPERT .LE. NPERT)
C
C-------PREPARE CALCULATION OF SIMULATION RESPONSE
        IF(GWMACTIVE)
     &      CALL GWM1RMS3PP(IOUT,IPERT,FIRSTSIM,LASTSIM,NPERT)
```

8—Immediately following the previous change, insert code to start new loop, point data to grid, retrieve name file, rewind files and reread BAS data before returning to all remaining GWF...AR calls

```
        DO IGRID = 1, NGRIDS
          CALL SGWF2BAS7PNT(IGRID)
          FNAME=FNAMEV(IGRID)
C GWM: PREPARE FLOW PROCESS FILES FOR NEXT SIMULATION
          IF(GWMACTIVE .AND. .NOT.FIRSTSIM)THEN
C---------REWIND FLOW PROCESS MODEL FILES AND RE-READ BAS FILE
            CALL GWM1BAS3RW(INUNIT,FNAME,CUNIT,NIUNIT,IOUT,
     &                  VERSION,MFVNAM)
            CALL GWM1BAS3RRF(INUNIT,CUNIT,VERSION,24,31,32,
     &                  IGRID,12,HEADNG,26,MFVNAM,FNAME,LASTSIM)
          ENDIF
```

9—Writing to screen at each time step interferes with GWM screen output, so comment out the lines near the beginning of the 90 time-step loop

```
C GWM: too much information to screen    WRITE(*,25)KPER,KSTP
C  25    FORMAT(' Solving:  Stress period: ',i5,4x,
C     &          'Time step: ',i5,4x,'Ground-Water Flow Eqn.')
```

10—Add GWM managed flows to FM block within GWF iteration loop

```
C GWM : ADD GWM MANAGED FLOWS
              IF(IUNIT(56).GT.0) CALL GWF2DCV3FM(KKPER,IGRID)
```

11—Add this code at the end of the LGR grid loop

```
C
C GWM: IF GWM ACTIVE TEST FOR CONVERGENCE FAILURE USING GWF CRITERIA
          IF(GWMACTIVE)THEN
            CALL GWM1BAS3AP(1,ILGR,ICNVG,LGRCNVG,IPERT)
            IF(.NOT.MFCNVRG(IPERT))THEN         ! SOLUTION NOT ACCEPTABLE
              CALL GWM1RMS3FP(IPERT,NPERT,FIRSTSIM,
     &                           LASTSIM)        ! RESET PERTURBATION PARAMETERS
              FIRSTSIM = .FALSE.
              GO TO 200                          ! SKIP TO PERTURBATION LOOP END
            ENDIF
          ENDIF
```

12—Add GWM managed flows to budget calculations within BD block

```
C GWM: RECORD BUDGET FOR GWM MANAGED FLOW VARIABLES
              IF(IUNIT(56).GT.0)CALL GWF2DCV3BD(KKSTP,KKPER,IGRID)
```

13—Insert test for convergence after budget calculations are complete

```
C
C GWM: IF GWM ACTIVE TEST FOR GWF FAILURE USING WATER BALANCE CRITERIA
            IF(GWMACTIVE)THEN
              CALL GWM1BAS3AP(2,ILGR,ICNVG,LGRCNVG,IPERT)
              IF(.NOT.MFCNVRG(IPERT))THEN        ! SOLUTION NOT ACCEPTABLE
                CALL GWM1RMS3FP(IPERT,NPERT,FIRSTSIM,
     &                             LASTSIM)        ! RESET PERTURBATION PARAMETERS
                FIRSTSIM = .FALSE.
                GO TO 200                          ! SKIP TO PERTURBATION LOOP END
              ENDIF
            ENDIF
```

14—Insert following lines after comment C7C6 and comment out existing control

```
C------GWM TAKES CONTROL OF LOOPING HERE
              IF(.NOT.GWMACTIVE.AND.ICNVG.EQ.0 .AND. ILGR .EQ. 0)GO TO 110
C                IF(ICNVG.EQ.0 .AND. ILGR .EQ. 0) GO TO 110
```

15—Insert following lines between 90 CONTINUE and 100 CONTINUE

```
C GWM:  ASSIGNS SYSTEM STATE TO STATE ARRAY AT END OF STRESS PERIOD
         IF(GWMACTIVE)THEN
           DO 95 IGRID=1,NGRIDS
             CALL SGWF2BAS7PNT(IGRID)
             CALL GWM1RMS3OS(KKPER,HDRY,IGRID) ! CHECK MANAGED WELLS FOR DEWATER
             CALL GWM1HDC3OS(IGRID,KKPER,IPERT,HDRY,0,DEPVALS)
             IF(IUNIT(18).GT.0)
     &         CALL GWM1STC3OS(IGRID,KKPER,0,0,DEPVALS)  ! STR PACKAGE
             IF(IUNIT(44).GT.0)
     &         CALL GWM1STC3OS2(IGRID,KKPER,0,0,DEPVALS) ! SFR PACKAGE
 95      ENDDO
```

16—Insert following lines immediately after 100 CONTINUE to complete simulation

```
IF(.NOT.GWMACTIVE)THEN ! GWM NOT ACTIVE
  IPERT = 2              ! INCREMENT IPERT TO TERMINATE PERTURBATION LOOP
ELSEIF(GWMACTIVE)THEN ! ANALYZE SIMULATION RESPONSE FOR GWM
  MFCNVRG(IPERT) = .TRUE.  ! SIMULATION HAS CONVERGED
  CALL GWM1RMS3FP(IPERT,NPERT,FIRSTSIM,LASTSIM)
  FIRSTSIM = .FALSE.      ! SIMULATION COMPLETED, NO LONGER FIRSTSIM
ENDIF
```

17—To prevent too much elapsed time between printed output, change call to GLO1BAS6ET

```
IF(LASTSIM) CALL GLO1BAS6ET(IOUT,IBDT,1)
```

18—Loop over all grids to perform GWF deallocations for all packages except BAS, LGR and solvers, and OBS Process packages. Insert code to do limited deallocation of BAS package

```
IF(.NOT. LASTSIM)  CALL GWM1BAS3DABAS7(IGRID)!DEALLOCATE SOME BAS7 MEMORY
```

19—Insert code to conclude looping, solve optimization problem, and test for GWM convergence

```
C GWM: END OF GWM PERTURBATION LOOP
  200   ENDDO
C
C-------IF NOT THE LAST SIMULATION, SOLVE GWM PROBLEM AT THIS ITERATION
       IF(.NOT. LASTSIM)THEN            ! GWM PROCESS MUST BE ACTIVE
           CALL GWM1RMS3FM              ! FORMULATE THE GWM PROBLEM
           CALL GWM1RMS3AP(GWMCNVRG)    ! SOLVE THE GWM PROBLEM
           IF(.NOT.GWMCNVRG)CALL GWM1RMS3OT(-1,NGRIDS)! WRITE GWM SOLUTION
       ENDIF
C
C-------IF GWM PROBLEM HAS CONVERGED, ONE MORE SIMULATION RUN IS NEEDED
       IF(GWMCNVRG .AND. .NOT. LASTSIM)THEN
           CALL GWM1RMS3OT(0,NGRIDS)    ! WRITE GWM SOLUTION OUTPUT
           LASTSIM = .TRUE.            ! EXECUTE ONE LAST SIMULATION
C
C-------IF GWM HAS COMPLETED ITS LAST SIMULATION OR GWM IS NOT ACTIVE
       ELSEIF(LASTSIM)THEN
           FINISH = .TRUE.            ! TERMINATE ITERATION LOOP
       ENDIF
C
C-----END OF GWM ITERATION LOOP
  300 ENDDO
```

20—Insert code to perform final deallocations before ending program

```
C
C GWM: DEALLOCATE REMAINING PACKAGES: LGR, SOLVER, OBS2 AND BAS7
C     GWF2BAS7DA MUST BE CALLED LAST BECAUSE IT DEALLOCATES IUNIT.
      DO IGRID = 1, NGRIDS
        CALL SGWF2BAS7PNT(IGRID)
        IF(ILGR.NE.0) CALL GWF2LGR1DA(IGRID)
        IF(IUNIT(9).GT.0) CALL SIP7DA(IGRID)
        IF(IUNIT(10).GT.0) CALL DE47DA(IGRID)
        IF(IUNIT(13).GT.0) CALL PCG7DA(IGRID)
        IF(IUNIT(42).GT.0) CALL GMG7DA(IGRID)
        CALL OBS2BAS7DA(IUNIT(28),IGRID)
        IF(IUNIT(33).GT.0) CALL OBS2DRN7DA(IGRID)
        IF(IUNIT(34).GT.0) CALL OBS2RIV7DA(IGRID)
        IF(IUNIT(35).GT.0) CALL OBS2GHB7DA(IGRID)
        IF(IUNIT(38).GT.0) CALL OBS2CHD7DA(IGRID)
        CALL GWF2BAS7DA(IGRID)
      ENDDO
```